Getting Started with Eclipse Juno

A fast paced tutorial to get you up and running with Eclipse Juno IDE

Rodrigo Fraxino Araujo
Vinicius H. S. Durelli
Rafael M. Teixeira

[PACKT] PUBLISHING open source community experience distilled

BIRMINGHAM - MUMBAI

Getting Started with Eclipse Juno

First published: July 2013

Production Reference: 1100713

Published by Packt Publishing Ltd.
Livery Place
35 Livery Street
Birmingham B3 2PB, UK.

ISBN 978-1-78216-094-6

www.packtpub.com

Cover Image by Siddhart Ravishankar (sidd.ravishankar@gmail.com)

Credits

Authors

Rodrigo Fraxino Araujo

Vinicius H. S. Durelli

Rafael M. Teixeira

Reviewers

Francisco Osorio Gálvez

Peter Rice

Acquisition Editors

Amarabha Banerjee

Rukhsana Khambatta

Lead Technical Editor

Ankita Shashi

Technical Editors

Sharvari H. Baet

Athira Lalji

Veena Pagare

Copy Editors

Insiya Morbiwala

Aditya Nair

Project Coordinator

Anugya Khurana

Proofreader

Clyde Jenkins

Indexer

Tejal R. Soni

Production Coordinator

Conidon Miranda

Cover Work

Conidon Miranda

About the Authors

Rodrigo Fraxino Araujo is a Ph.D. candidate in Computer Science at the University of São Paulo, Brazil. He has spent a year as a visiting scholar at the Institut National de Recherche en Informatique et Automatique, Rocquencourt, France. He is also a software engineer and since 2011, is working at IBM.

I would specially like to thank Carolina, who gave me the strength to work on this book.

Vinicius H. S. Durelli is a Ph.D. candidate in Computer Science at the University of São Paulo, Brazil. As part of his doctoral studies, from 2011 to 2012, he was a visiting scholar at the George Mason University, Virginia, USA. He received his M.S. in Computer Science from the Federal University of São Carlos in 2008. He has been a Sun Certified Java Programmer since 2006. Also, he has been using Eclipse since circa 2004, which makes him feel old. When he is not writing or programming, Vinicius enjoys playing video games (especially Mario and Zelda games) and practicing Brazilian Jiu Jitsu.

I would like to express my deepest gratitude to my friend Simone Borges for placing food in front of me in order to keep me going while I was trying to hit deadlines related to this book and my ongoing Ph.D. studies.

Rafael M. Teixeira currently works as a Software Engineer at IBM Linux Technology Center, developing code for Eclipse open source projects. He's currently taking a MSc program in University of São Paulo, where he also received his Computer Engineering degree. Rafael's favorite hobby is running, but also enjoys some occasional video gaming.

I would like to thank my family for their caring and support, which made everything possible. I would also like to thank Bianca Teixeira, whose motivating and sweet words have kept me going for the last 7 years and will hopefully do so for many others. I dedicate this work to the memory of Luis Fernando Teixeira.

About the Reviewers

Francisco Osorio Gálvez is a senior developer and technical leader working at Citigroup, and previously at IBM. He works designing, developing, and leading large Java enterprise systems for the banking industry. He enjoys working on different technologies just for fun. You can contact him at `fosoriog@gmail.com`.

> I would like to thank my wife Fabiola for the time I stole from her. Weekends were sacrificed in the review of this book. I'll make it up to you, I promise. ;)

Peter Rice is Professor Emeritus of Mathematics from the University of Georgia, USA. He began a career in the IT industry after retirement, which led him to becoming a certified Microsoft trainer, an IBM trainer, and to do extensive work as an independent trainer working from his own company, PM Rice Consulting, Inc. In addition, he has created and worked as principal architect on a number of commercial programs. He also works in a number of languages.

Table of Contents

Preface **1**

Chapter 1: Introduction **9**

How to use this book **9**
Downloading and installing Eclipse **10**
Eclipse Juno – new and noteworthy **11**
Summary **18**

Chapter 2: Java Development **19**

Creating a Java project **20**
Creating a Java class 21
Creating working sets 21
Importing a Java project **22**
Importing a project from Version Control Servers 23
Introducing Java views **23**
The Package Explorer view 23
The Java Editor view 24
Compiling errors and warnings annotations 25
Content assist 25
Code navigation 26
Quick fix 27
Customizing the editor 29
The Problems view 30
Customizing the Problems view 30
The Outline view 31
The Type Hierarchy view 32
The Call Hierarchy view 32
Organizing imports 33
Save actions 33
Enforcing Coding Style with Formatter 34

Generating code **34**
Generating getters and setters 34
Generating constructors 35
Generating the hashCode() and equals() methods 36
Generating the toString() method 36
Generating method comments 36
Editing code and comment templates 37
Refactoring **38**
Renaming variables 38
Modifying a method's signature 39
Building and running the project **39**
Creating a launch configuration 40
Managing the project build path **41**
Summary **42**
Chapter 3: Unit Testing with JUnit and Debugging **43**
Testing with JUnit – getting started **44**
Setting up JUnit 46
Testing with JUnit 49
Implementing the test methods generated by Eclipse 53
An overview of the Eclipse JUnit Runner view 57
Creating test cases 58
Debugging when problems arise **67**
Summary **72**
Chapter 4: Version Control Systems **73**
CVS **74**
Repository 74
Checking out a project 76
Synchronizing 76
Branches 76
SVN **77**
Git **80**
Summary **86**
Chapter 5: SWT **87**
Getting started **88**
Setting up 88
Widgets 90
Displays 92
Controls **92**
The Label widgets 92
The Text widgets 94

The Button widgets 96
The List widgets 97
The Combo widgets 99
Layouts **100**
FillLayout 101
RowLayout 102
GridLayout 104
New and noteworthy **107**
Summary **107**
Chapter 6: More SWT **109**
Events **109**
Typed and untyped events **110**
KeyEvent 112
MouseEvent 114
Menus **115**
Toolbars **117**
Tables **118**
Dialogs **121**
MessageBox 121
FileDialog 123
ColorDialog 124
FontDialog 125
PrintDialog 126
Summary **126**
Chapter 7: Web Development Using Eclipse WTP **127**
Brief introduction to the benefits of Java web application technologies **128**
Understanding servlets **129**
Understanding JavaServer Page **129**
Getting started with Apache Tomcat **130**
Installing Tomcat **130**
Eclipse Web Tools Platform **134**
Configuring the server 137
The servlet lifecycle **140**
Implementing your very first Java servlet using Eclipse WTP **141**
A more elaborate example – yet another calculator 151
Deploying the calculator example 154
Implementing your very first JSP using Eclipse WTP **160**
Generating WAR files **164**
Summary **166**

Chapter 8: Eclipse Development **167**
Creating your first plugin **168**
Running and debugging a plugin project **171**
Running and debugging configurations 171
Extension points **173**
Declaring an extension point 173
Providing interfaces 176
Using extension points implementation in code 177
Implementing an extension point 179
Contributing to the platform's menus and toolbars **180**
Actions versus commands 181
org.eclipse.ui.menus 181
org.eclipse.ui.commands 182
org.eclipse.ui.handlers 182
Restricting and disabling contributions 184
The Plugin Spy feature 188
Creating new views 190
Saving the view's current state 191
Adding context help to your view 191
Example of a new view 192
Exporting a plugin **194**
Summary **195**
Chapter 9: Eclipse Rich Client Platform **197**
Understanding a Rich Client Platform **198**
The Eclipse Rich Client Platform **198**
OSGi framework implementation 199
The module layer 199
The life cycle layer 199
The service layer 200
The security layer 200
SWT 200
JFace 200
Eclipse workbench 201
Developing a client application using the Eclipse RCP **201**
Creating a new client application project using a template 202
A closer look at the generated source code 203
Running and debugging RCP applications **207**
Expanding the example – creating a contact list application **208**
The contact list view 210
The contact editor 213
The Save and New commands and menu entries 217
Tying up the two views 219
Running the application 221

Packaging and branding an Eclipse RCP application **222**
Branding an Eclipse RCP application 224
Summary **226**
Appendix: Keyboard Shortcuts **227**
File editor shortcuts **227**
Code edition shortcuts 227
Code generation and code refactoring shortcuts 228
Code navigation shortcuts 229
Java shortcuts 229
File shortcuts **230**
Run and debug shortcuts **230**
Views shortcuts **231**
Index **233**

Preface

The first version of Eclipse was released in the November of 2001. Since then, it has come a long way in terms of enhancements and number of users. Due to the many improvements it has undergone, over time Eclipse has become popular and received a great deal of praise. Nowadays, Eclipse is arguably the most widely used integrated development environment (IDE) for developing Java applications. It is free and open source, and is shipped with plugins for performing lots of software-development-related activities. In addition, more plugins can be easily downloaded and installed according to your personal needs and preferences. Another plus is that, since Eclipse is written mostly in Java, it is cross-platform.

If you have been using Java without the help of an IDE, you must have realized by now that Java can be rough and unnerving at times. Most of the time, the compiler (javac) does not display helpful error messages. Even something as simple as an omitted semicolon may cause it to display a lengthy list of error messages. Inside an IDE, this sort of problem can be spotted before you try to run the program. An IDE can even suggest solutions before you start scratching your head. Undoubtedly, using a full-fledged IDE provides a range of benefits. For instance, an IDE provides autocompletion of your code, it helps you to format your code properly, and some of them come with nifty plugins for testing, debugging, and keeping track of the changes that you make to your code.

However, despite all the benefits involved in using an IDE, let's be upfront; learning how to use one of these high-end development tools is no bed of roses. IDEs have a steep learning curve, which means that mastering such a tool usually takes time and effort. To make matters worse, sometimes, many developers try to learn a new programming language and an IDE in tandem.

This book will help you to overcome Eclipse's steep learning curve. After reading this book, you will be grounded in performing many development activities using Eclipse.

What this book covers

In this book, we cover Eclipse Juno that is, at of the time of this writing, the latest stable version of the IDE (it was released in June 2012). The book starts off by showing how to perform the most basic activities related to implementing Java applications (such as creating and organizing Java projects, refactoring, and setting launch configurations), working up to more sophisticated topics as testing, web development, GUI programming, and extending Eclipse's editors to suit your needs.

This book takes a hands-on approach, diving straight into using Eclipse without getting you bogged down in technical details. Here are some of the topics covered in this book:

- Effectively using Eclipse to write Java code
- Integrating JUnit 4 into Eclipse, develop, and run JUnit-based test methods
- Managing your project using the version control system Git/EGit
- Developing GUI applications using SWT
- Advanced GUI development with examples
- Using Eclipse WTP plugin to develop and deploy web-based Java applications
- Extending Eclipse by developing new editors, views, and contributing to existing elements of the IDE

Chapter 1, Introduction, covers how to get started with Eclipse. It goes from how to get Eclipse up and running to a brief overview of the new and noteworthy features of Eclipse Juno.

Chapter 2, Java Development, covers basic Java development using Eclipse. To beginners, Eclipse can be quite intimidating. The main window contains lots of smaller windows and some of the smaller windows have a bunch of buttons and drop-down menus. This chapter is all about getting to know the main features of Eclipse for developing Java programs. We will go through all the steps to get you going from creating Java projects, detailing the Eclipse views used for Java development, Java code generation, to building and running Java projects, which involves setting up launch configurations. Also, we cover some refactorings, so that you will be able to improve your Java code – as you tackle more complicated projects will you see that refactoring is a very important aspect of Java development.

Chapter 3, Unit Testing with JUnit and Debugging, introduces JUnit, the most widely used framework for implementing unit tests in Java. You will learn how to set JUnit up so that it can be used in conjunction with Eclipse. In addition, this chapter describes the creation of test methods using JUnit's annotations. As you will see, annotations can also be used to inform JUnit when to skip a particular test method and whether or not a given method is expected to throw an exception. The chapter wraps up with an introduction on how to use Eclipse's debugger to pinpoint tough to find problems in your code.

Chapter 4, Version Control Systems, explains using Eclipse in conjunction with version control systems. When working on a complex software project (or one that involves more than one person), you need to use a tool that manages and tracks different version of the software project in question. This book covers CVS, SVN, and Git, and how to use it inside Eclipse through native and external plugins. After reading this chapter, you will be able to use them inside Eclipse to support collaborative development.

Chapter 5, SWT, covers graphical user interface (GUI) development with the Standard Widget Toolkit (SWT). This chapter introduces the most basic widgets as, for example, buttons, labels, and combos. In a way, this chapter lays the foundation for *Chapter 6, More SWT*.

Chapter 6, More SWT, builds on top of the previous chapter, showing how to make more appealing GUIs by introducing advanced widgets as, for example, toolbars and tables. This chapter also features a full-blown example.

Chapter 7, Web Development Using Eclipse WTP, is concerned with using Eclipse to develop basic web applications. Throughout this chapter we show how Eclipse, along with a set of plugins called Web Tools Platform (WTP), can be used to support the development of web applications that are based on two basic web technologies, namely, servlets and JavaServer Pages (JSP).

Chapter 8, Eclipse Development, covers a more advanced topic: extending Eclipse. You will learn how Eclipse's architecture is organized. Further, by reading this chapter, you will see how extension points can be harnessed in order to create your own plugin, editor, or even extend a particular view.

Chapter 9, Eclipse Rich Client Platform, shows how you can leverage what you have learned in the previous chapter to develop your own rich client application using the Eclipse Rich Client Platform. By creating a sample application, you will learn how to add views, menus, and panels, as well as make them interact with each other. We will also cover how to brand and package your application, allowing you to build a fully functional, cross-platform application using the Eclipse platform.

Appendix, Keyboard Shortcuts, contains all keyboard shortcuts presented throughout the book, plus other useful ones that can greatly improve your productivity with Eclipse.

What you need for this book

The two pieces of software needed for this book are a Java Virtual Machine (JVM) and Eclipse. Both are free and available for download. Some additional plugins and software are also required, but we will detail how to get and install them along the book. The next section details how to install Eclipse and provides some pointers on where to find information concerning how to install a version of the JVM.

Who this book is for

Although Eclipse can be used as IDE for a number of different languages, it is more popular within the Java community. We wrote this book with a Java programmer interested in getting the most out of Eclipse in mind. Therefore, the main focus is going to be Java development using Eclipse, and the primary audience for this book is the Java programmer.

The book was written in a way that it is accessible both to beginners and intermediate level Java programmers. This book will also be beneficial to a graduate student who is familiar with Java. Practicing software developers might use this book as their first foray into the Eclipse world. If you are a seasoned Java developer that has been using another IDE and wondering what Eclipse brings to the table, this book will provide you with a hands-on walkthrough of the main IDE features. Seasoned Eclipse users might not benefit from this book, since it is tailored to be an introduction on how to get started with Eclipse. Let's put it another way, this book can be seen as a first step towards mastering Eclipse; it is a fast-paced tutorial for those who want to hit the ground running and begin being productive with Eclipse right away.

If you are not familiar with Java, then this is not the book for you. If you have just been introduced to Java, you will be able to follow most of the content in this book after a crash course in Java syntax (just make sure you have a Java textbook close at hand.). For the most part, the chunks of Java code presented here are accessible to Java beginners. We tried to keep the code in the examples as si mple as possible so that beginners will not be bogged down by technical details related to Java syntax. Whenever something a little bit more complex is shown, we provide an overview of it or we direct the reader to a more in-depth discussion of the to pic at hand.

Conventions

In this book, you will find a number of styles of text that distinguish between different kinds of information. Here are some examples of these styles, and an explanation of their meaning.

Code words in text are shown as follows: " The option to transform a `for` loop into an indexed `for` loop is available for expressions."

A block of code is set as follows:

```
@Test(timeout=1000)
public void testSquareRoot() {
   while(true);
}
```

When we wish to draw your attention to a particular part of a code block, the relevant lines or items are set in bold:

```
@Test(timeout=1000)
public void testSquareRoot() {
   while(true);
}
```

New terms and **important words** are shown in bold. Words that you see on the screen, in menus or dialog boxes for example, appear in the text like this: " We strongly recommend you to download the **Eclipse IDE for Java Developers**.

> Warnings or important notes appear in a box like this.

> Tips and tricks appear like this.

Reader feedback

Feedback from our readers is always welcome. Let us know what you think about this book—what you liked or may have disliked. Reader feedback is important for us to develop titles that you really get the most out of.

To send us general feedback, simply send an e-mail to `feedback@packtpub.com`, and mention the book title via the subject of your message.

If there is a book that you need and would like to see us publish, please send us a note in the **SUGGEST A TITLE** form on `www.packtpub.com` or e-mail `suggest@packtpub.com`.

If there is a topic that you have expertise in and you are interested in either writing or contributing to a book, see our author guide on `www.packtpub.com/authors`.

Customer support

Now that you are the proud owner of a Packt book, we have a number of things to help you to get the most from your purchase.

Downloading the example code

You can download the example code files for all Packt books you have purchased from your account at `http://www.PacktPub.com`. If you purchased this book elsewhere, you can visit `http://www.PacktPub.com/support` and register to have the files e-mailed directly to you.

Errata

Although we have taken every care to ensure the accuracy of our content, mistakes do happen. If you find a mistake in one of our books—maybe a mistake in the text or the code—we would be grateful if you would report this to us. By doing so, you can save other readers from frustration and help us improve subsequent versions of this book. If you find any errata, please report them by visiting `http://www.packtpub.com/support`, selecting your book, clicking on the **errata submission form** link, and entering the details of your errata. Once your errata are verified, your submission will be accepted and the errata will be uploaded on our website, or added to any list of existing errata, under the Errata section of that title. Any existing errata can be viewed by selecting your title from `http://www.packtpub.com/support`.

Piracy

Piracy of copyright material on the Internet is an ongoing problem across all media. At Packt, we take the protection of our copyright and licenses very seriously. If you come across any illegal copies of our works, in any form, on the Internet, please provide us with the location address or website name immediately so that we can pursue a remedy.

Please contact us at `copyright@packtpub.com` with a link to the suspected pirated material.

We appreciate your help in protecting our authors, and our ability to bring you valuable content.

Questions

You can contact us at `questions@packtpub.com` if you are having a problem with any aspect of the book, and we will do our best to address it.

1
Introduction

This chapter sets the stage for the next ones. Here we cover how to get Eclipse up and running. Further, we also briefly describe what is new in this new version of the **Integrated Development Environment** (IDE). If you do not have any interest in knowing what new features have been added to Eclipse, you can safely skip this chapter.

We also cover how you should approach this book, where we answer questions, such as "Do I need to read the chapters in the sequence in which they appear in the book?" and "This book is riddled with tip boxes; do I need to read each and every one of them?".

Summing up, the contents of this chapter are as follows:

- Downloading Eclipse
- Installing Eclipse
- New and noteworthy features in Eclipse Juno

How to use this book

As mentioned, we assume that you know Java (we expect you to know at least the basics) and, most importantly, that you have a working knowledge of computers. After all, we will not cover the nitty-gritty details of how to unpack files, create directories, find a file, start a program, and configure environment variables.

As you will see, in each chapter there are boxes containing tips and pitfalls regarding Eclipse, Java, and, sometimes, the topic at hand. These boxes are meant to help you get the most out of the underlying chapter's content, so do not skip them.

Although this book follows a tutorial-like approach, you do not need to read the chapters in the sequence in which they appear in the book. All chapters are self contained. You can read the chapters in this book in any particular order.

Downloading and installing Eclipse

Eclipse is a Java program; this means that in order to use Eclipse, you first need to install a version of the **Java Virtual Machine** (**JVM**). There are basically two ways of achieving this: you may either install a **Java Runtime Environment** (**JRE**) or a **Java Development Kit** (**JDK**). Actually, what you should install boils down to what you intend to accomplish with Eclipse. Since you want to use Eclipse for Java development, you should install a JDK (besides the JVM, the JDK includes other useful things; for instance, the source code for the standard Java libraries). To run Eclipse Juno, it is recommended that you install JDK 6 or a later version (many Eclipse users have been using the newer version—Java 7). To learn how to install a version of the JVM on your operating system, navigate to `http://www.oracle.com/technetwork/java/javase/index-137561.html`.

If you have already installed a JVM, downloading and installing Eclipse will be quite straightforward:

1. First, download Eclipse Juno from the Eclipse downloads page at `http://www.eclipse.org/downloads/`. We strongly recommend you to download the **Eclipse IDE for Java Developers** version. Also, make sure to download the right version for your platform and Java version (there are two versions: 32 bit and 64 bit). Actually, the download site is smart enough to detect the OS you are running, so it will default to the download options that are available to your platform. You will get Eclipse as a compressed file (it might be either a `.zip` or `.tar.gz` file). It is worth mentioning that there are several distributions of the IDE; each of these distributions is tailored towards a particular purpose. Therefore, the set of plugins that ships with each distribution varies. Some distributions are Eclipse IDE for Java EE Developers, Eclipse for Mobile Developers, Eclipse Modeling Tools, and Eclipse for Testers.
2. Second, unpack this file into the directory of your choice. As a result, you will get a subdirectory named `eclipse` containing lots of subdirectories under it. After unpacking it, you are good to go. No further configuration is required. Easy, isn't it?

> There is a known issue with the built-in decompression utility shipped with the current versions of Windows. Therefore, you had better use another decompression utility when decompressing Eclipse. If you are a Windows user looking for a more robust decompression utility, give 7-Zip a try (`http://www.7-zip.org/download.html`). In addition, some people have reported that Eclipse has a problem with directory paths that contain spaces in their names. So if possible, try to avoid extracting Eclipse in a directory path containing spaces in its name.

To start Eclipse, double-click on the file containing Eclipse's icon (the file is named `eclipse.exe` on Windows and `eclipse` on Linux and Mac systems). Eclipse will prompt you to select a workspace. Within this context and in Eclipse's parlance, a workspace is the directory wherein your Eclipse projects will be placed. You had better create a new directory and name it `eclipse_workspace` (or something along the line) and point Eclipse to the location of your recently created directory. However, it bears mentioning that you can use any directory as an Eclipse workspace. It is also possible to enter the directory name directly in the workspace prompt, and Eclipse will take care of creating it for you.

After installing Eclipse, you are already able to beef it up by adding plugins of your choice. To do so, navigate to **Help | Install new software...**. We will cover plugins' installation in later chapters. For the time being, you just need to keep in mind that Eclipse can be extended right off the bat.

In the next section, we will highlight some of the changes that Eclipse has undergone. Of course, a lot has changed. However, let's not get bogged down with technical details. So in the next section, we will concentrate on describing only the most important and useful improvements (you will have to trust us on this one). If you are one of those people who crave completeness, you will be glad to know that we will also mention where you can get the complete list of improvements.

Eclipse Juno – new and noteworthy

Two flavors of Eclipse Juno have been included in this release: 4.2 and 3.8. The basic difference between these two flavors is that 4.2 is built on top of the e4 (Eclipse 4) platform and has undergone a lot of GUI changes, but 3.8 is based on Version 3 of the platform. It is also fair to say that the 3.8 release focuses on stability and bug fixes, whereas Eclipse Juno 4.2 emphasizes the introduction of new features.

The default appearance of the platform's user interface has been revamped. The main goal of this visual update was to adopt a more modern visual style. Eclipse Juno features a workbench with a whole new look, as shown in the following screenshot. To better understand this screenshot and how things are visually organized in Eclipse, we need to go through some concepts. The term workbench refers to the desktop development environment. So, as shown in the screenshot, the workbench encompasses pretty much everything else. The elements inside the workbench are grouped into perspectives, in Eclipse's parlance (the surrounded area indicated by 2 in the screenshot). Perspectives, in turn, contain views and editors (in the following screenshot, the letter *a* represents views, whereas the letter *b* represents editors). They control what appears in the workbench. Each perspective provides all you need to perform a certain task. For instance, the **Debug** perspective contains the set of functionalities you would use while debugging Java programs. This functionality is split into views.

An example of a view that belongs to the **Debug** perspective is the **Breakpoint** view. (If you had never heard about breakpoints before this, do not worry. We will cover everything about them in a later chapter.) As for editors, most perspectives contain several views and a single editor. In Eclipse, as many editors as you want can be open at a given time. However, only one can be active at a time. As you progress through the book, you will switch perspectives very often. Also, you will learn how to use the most important views and editors.

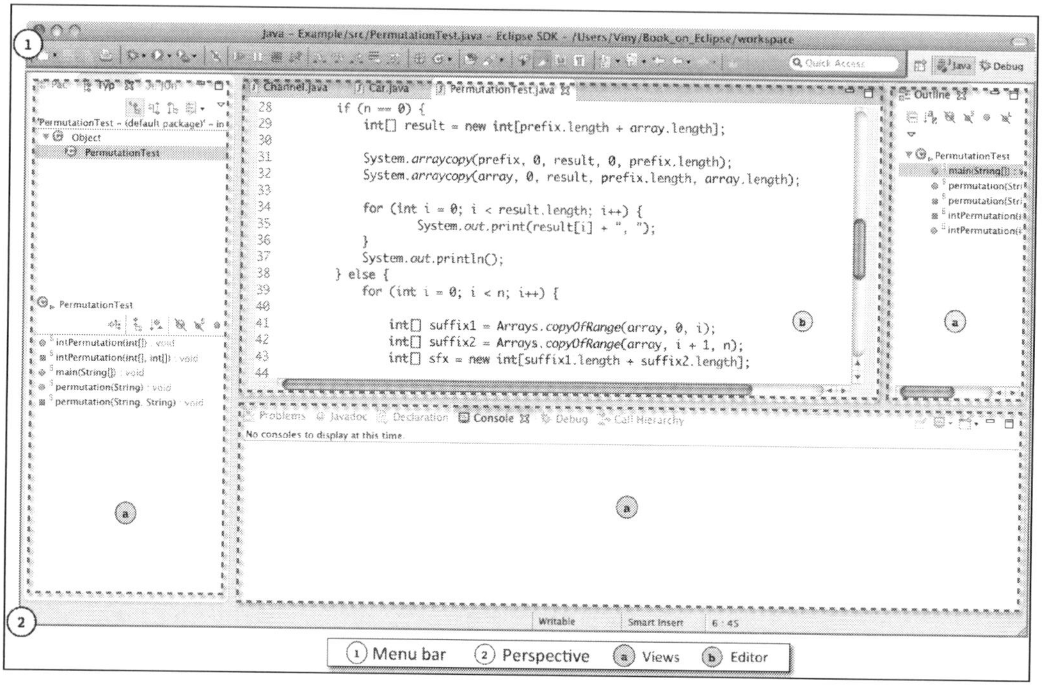

At a glance, the two changes that will certainly draw the attention of the reader familiar with Eclipse are the debug control commands (`Resume`, `Stop`, `Step Into`, and so on). Since these have been moved out of the **Debug** view into the global toolbar (we will cover debugging in *Chapter 3, Unit Testing with JUnit and Debugging*), we now have a global search bar in the top-right corner. These elements can be seen in the previous screenshot in the area indicated by the number 1 (that is, on the menu bar or the global toolbar).

Have you ever wanted to use a view side by side with an editor? Now, in Eclipse Juno, it is also possible to stack views and editors. For instance, if you have a view whose information when displayed takes up too much space, you can stack it in the editor area to provide more space. To do so, just drag the view by its title bar while keeping the left button of the mouse pressed. As you move the view around the workbench, the mouse pointer will indicate where and what shape the view will take if you release the left mouse button. When the view is in the location that you want (in this case, between two editor tabs), release the left mouse button. The following screenshot shows a view (the **Data Source Explorer** view) that was docked in the editor area. You can do the same with any view. It is also worth mentioning that you can place views pretty much wherever you want.

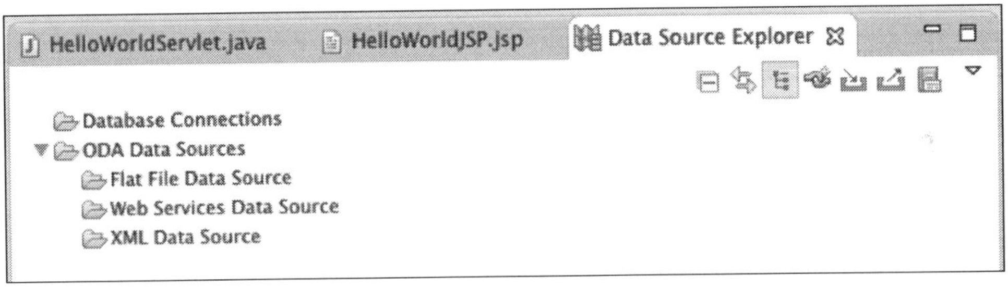

Another useful improvement is that editors can now be detached into their own windows. And what is better, the detached editor will retain its properties; this means that it will behave in exactly the same way as an editor still attached to the workbench window. This improvement is shown in the following screenshot. To detach an editor from Eclipse, simply drag it out of Eclipse's workbench window (main window).

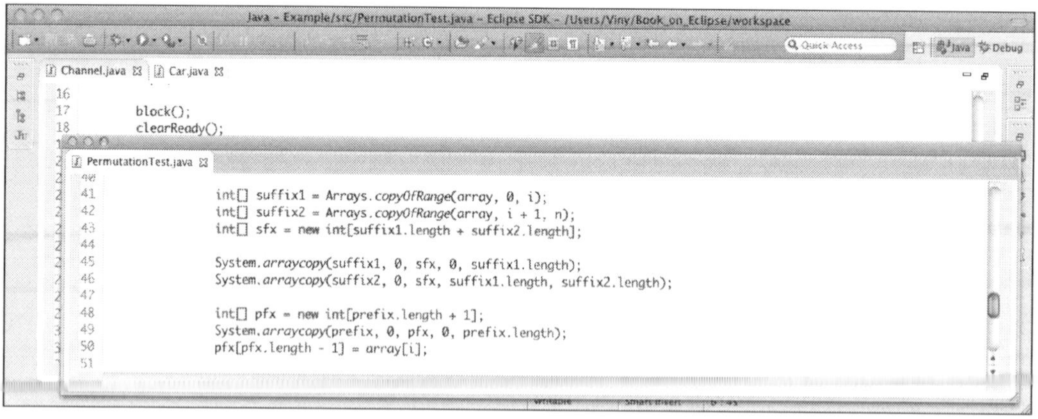

If you are one of those developers who try to get as much done as possible without lifting your fingers from the keyboard, you will find the new global search bar interesting and useful (the global search bar is in the top-right corner of the workbench). It allows you to access classes, views, perspectives, commands, menus, and so on quickly just by typing their names. The following screenshot shows the global search bar in action. To access the search bar without using the mouse, press *Ctrl* + *3*.

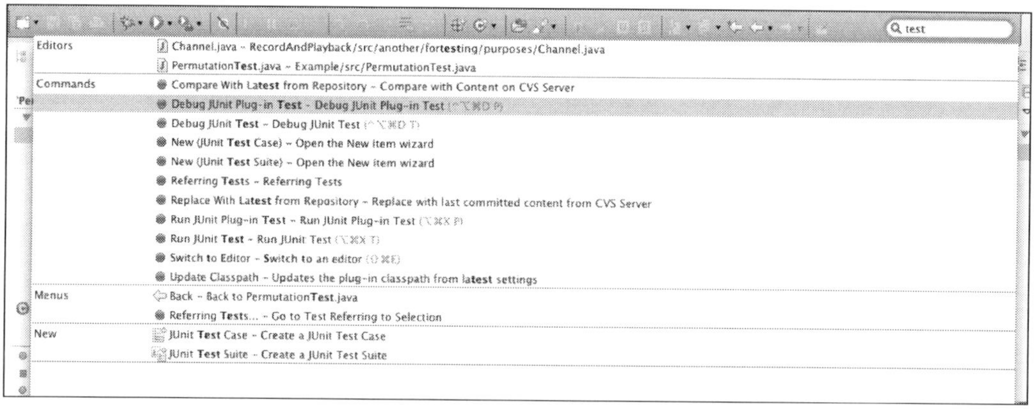

In the previous versions of Eclipse, a lightweight refresh mechanism was introduced. This mechanism keeps the contents of the workspace synchronized with the filesystem; the out-of-sync files are automatically, asynchronously refreshed. However, in previous versions of the IDE, you had to enable this feature.

Now, however, it is enabled by default. If you want to disable it, navigate to **Preferences | General | Workspace,** and deselect **Refresh on access**. Unless you have a very strong reason to do so, you should keep it enabled.

Part of the **Debug** perspective has been moved to the main menu bar. To be more specific, the common debug control commands (that is, `Resume`, `Stop`, and so on) have been moved to the main menu bar (as shown in the following screenshot). If you do not like it this way, you can customize it. Navigate to **Window | Customize Perspective...**. Here, you can either add more debugging commands or remove them from the main menu bar altogether.

> If you are feeling a little bit nostalgic and want everything as it used to be, that is, without the debugging command all over the place, that old look can be restored. To do so, first toggle the **Show Debug Toolbar** option in the **Debug** view's menu, and then disable the Debug toolbar from the **Customize Perspective** dialog. Also, if you want to return to the classic look (preserving the new features), you can navigate to **Preferences | General | Appearance** and select **Classic** from **Theme**.

Eclipse Juno ships with improved launch configuration support. From now on, if you'll create a launch configuration with the same name as that of an existing launch configuration, the type of conflicting launch configuration will be shown in the error message. The following screenshot shows the error message that you will get when trying to create a launch configuration with the same name as a previous one. (You will learn how to create launch configurations for your Java projects in the next chapter.)

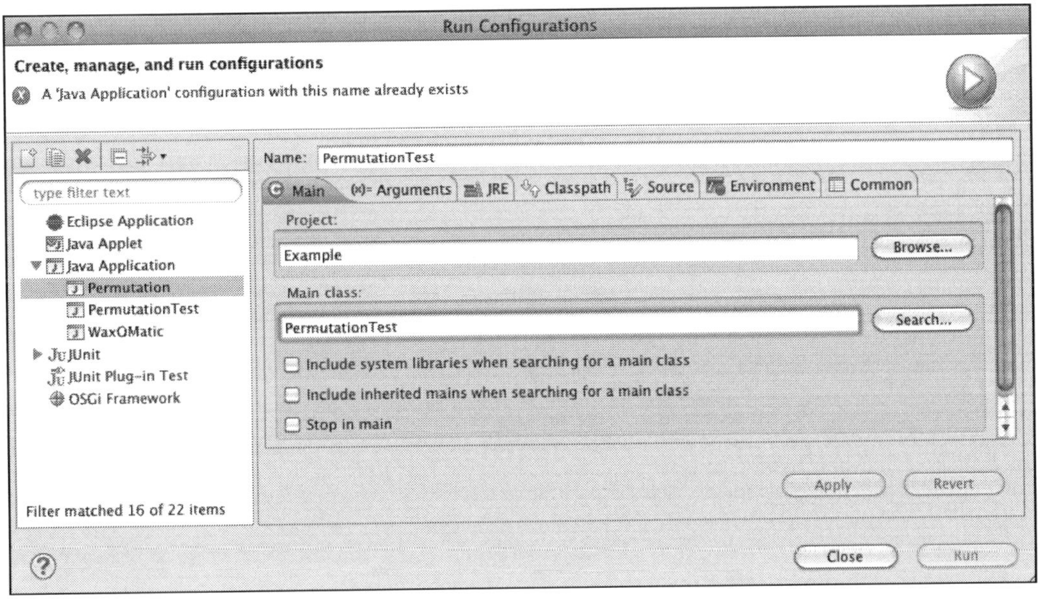

If you are already familiar with Eclipse, you have probably used the **Quick Outline** (*Ctrl* + *O*) and the **Quick Hierarchy** (*Ctrl* + *T*) views. These useful time-saving views are even better in Eclipse Juno: they now support CamelCase matching on the element name. For example, to search for ByteArrayOutputStream, simply type `BAOS`. The following screenshot shows a usage example:

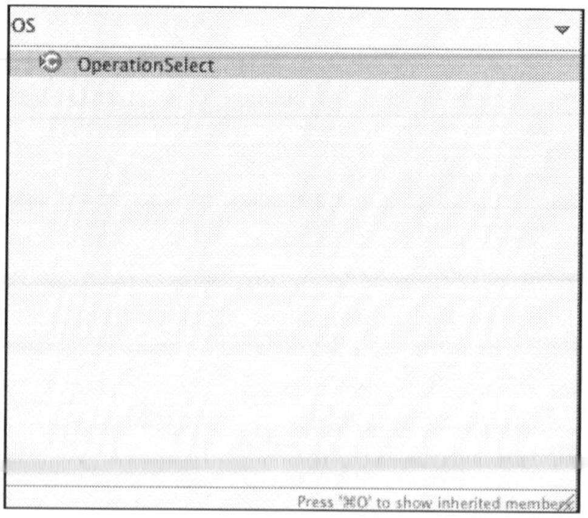

> If you are a Mac OS X user, use the command button (□) to access the **Quick Outline** and **Quick Hierarchy** views instead of *Ctrl*.

Another useful thing is the improved bracket-matching support in the Java editor. Now this editor supports the highlighting of enclosed brackets. This can be configured by going to the preference page at **Java** | **Editor** and then selecting the option **Enclosing brackets,** as shown in the following screenshot:

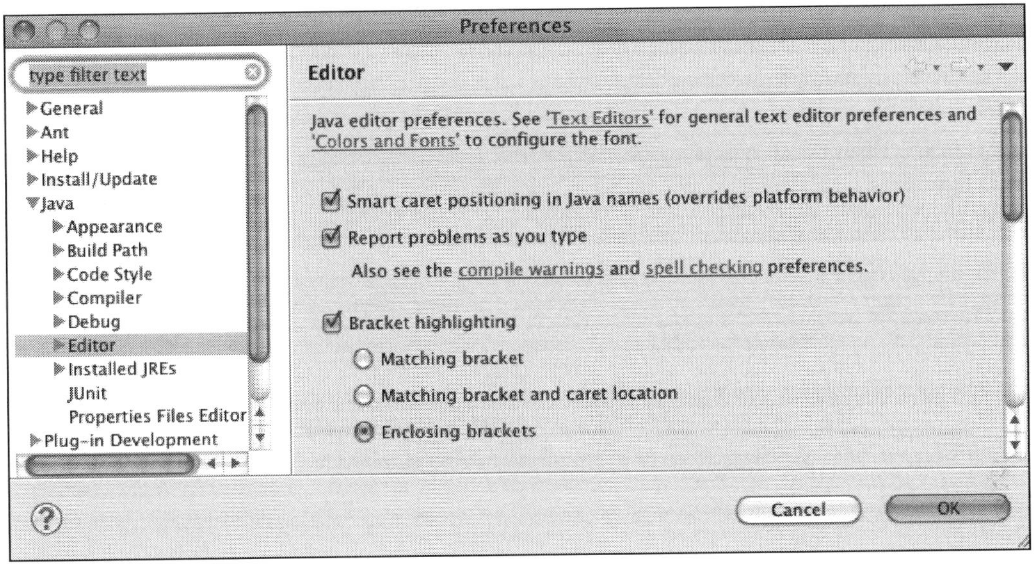

After enabling this option, go to the Navigate menu end select **Go To | Matching Bracket** (**Ctrl + Shift + P**), now this option works everywhere in the Java files. When inside a pair of brackets, pressing *Ctrl* + *Shift* + *P* navigates the caret to the nearest enclosing end bracket. On pressing the same buttons a second time, the caret navigates to the corresponding start bracket. Further, when you hover over a closing curly brace, Eclipse shows you the source code near the matching opening brace (as highlighted in the following screenshot).

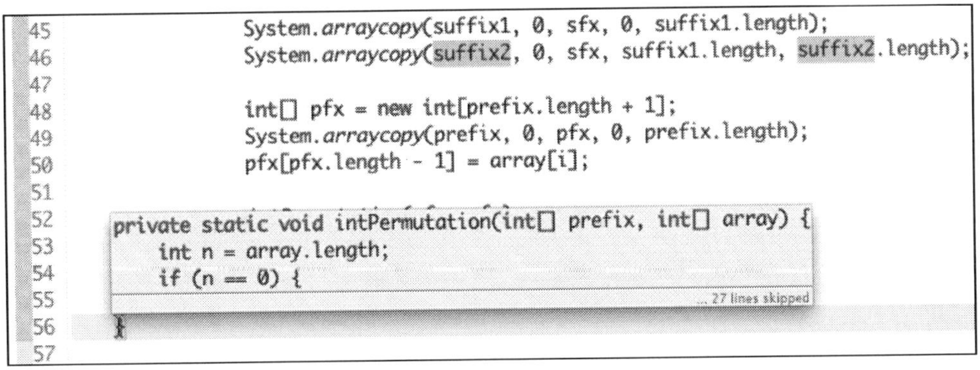

There is way more than we described here. However, giving an in-depth overview of every improvement will not buy you anything. You'd better start on your journey towards Eclipse mastery. All the skin-deep improvements Eclipse has undergone are documented at `http://archive.eclipse.org/eclipse/downloads/drops4/R-4.2-201206081400/news/eclipse-news-part1.html`.

A lot of non-GUI-related improvements have also been made. For instance, now you can use the quick assist (*Ctrl* + *1*) to turn an enhanced `for` loop into either an indexed `for` loop or an iterator-based loop. The option to transform a `for` loop into an indexed `for` loop is available for expressions that iterate over arrays or list-based types. The option to turn a `for` loop into an iterator-based `for` loop is only available for expressions of the type `iterable`. The following screenshot shows the new quick assist in action:

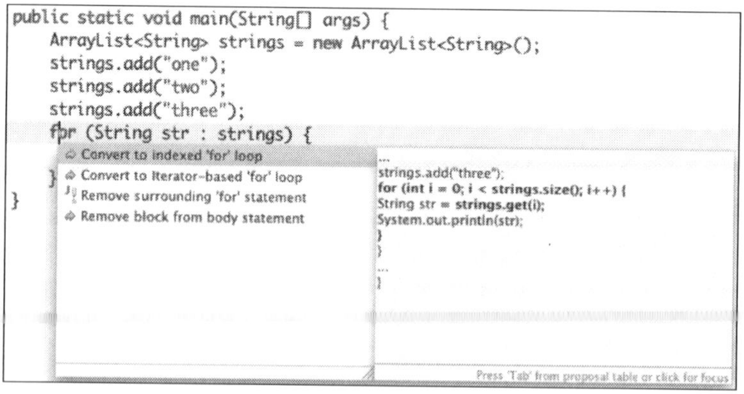

As we will see in *Chapter 3, Unit Testing with JUnit and Debugging*, Eclipse now ships with JUnit 4.10. Also, Eclipse supports the latest servlet application programming interface (3.0) that we will briefly discuss in *Chapter 7, Web Development Using Eclipse WTP*. The complete list of improvements is available at `http://help.eclipse.org/juno/index.jsp?topic=%2Forg.eclipse.platform.doc.user%2FwhatsNew%2Fplatform_whatsnew.html`.

Summary

Eclipse is arguably the most popular Java IDE (maybe that is why you bought this book; you figured it was about time you jumped on the bandwagon). In this chapter, we introduced Eclipse Juno and what this new version of the IDE brings to the table. As you saw, Eclipse has undergone lots of changes.

Besides introducing Eclipse, this chapter laid the foundation for the rest of the book. We emphasized some of the changes of the new version of the IDE. We could have spent more pages describing what is new, but that would not be appropriate for those of you who can barely wait to start using Eclipse. Above all, the most important changes have already been documented elsewhere (as mentioned before).

The next chapter delves into our hands-on tutorial, describing how to use Eclipse to develop Java applications. Good reading, and make sure not to get sidetracked by social networking sites.

2

Java Development

IDEs have evolved so much along the years that it's hard to imagine writing Java code without using one. It's always possible, of course, to use text editors or very simple IDEs, but Eclipse provides so many features to make the writing of code easier and faster, that once you get used to it, chances are you will never look back.

Given the productivity gain that an IDE like Eclipse might provide, mastering it is almost as important as mastering Java programming itself. The goal of this chapter is to present the most important features Eclipse has to offer to Java programmers.

By the end of this chapter, you will be able to:

- Create a new Java project or import an existing one
- Navigate swiftly through the code
- Use Java views effectively
- Refactor code in an easy and safe way
- Get Eclipse to generate code and fix code issues

Creating a Java project

To create a new Java project, navigate to **File** | **New** | **Project**. You will be presented with the **New Project** wizard window that is shown in the following screenshot:

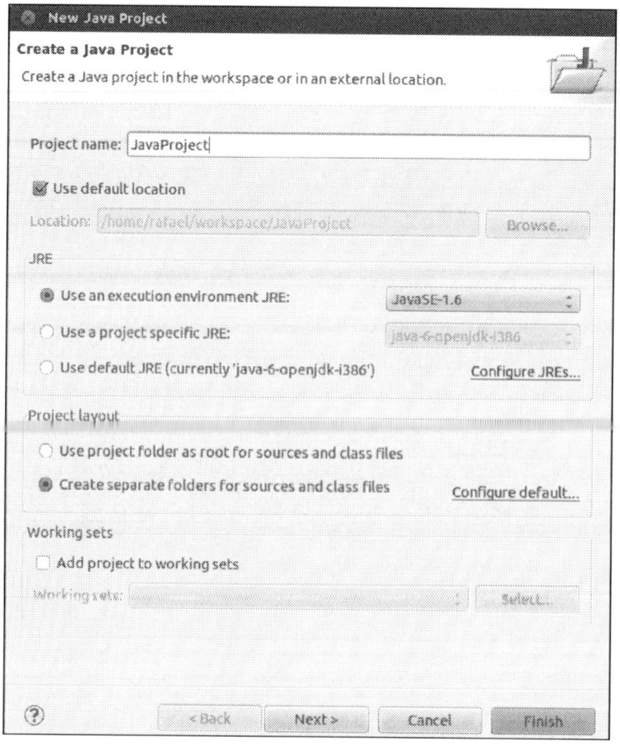

Choose the **Java Project** option, and click on **Next**. The next page of the wizard contains the basic configuration of the project that you will create. The **JRE** section allows you to use a specific JRE to compile and run your project. The **Project layout** section allows you to choose if both source and binary files are created in the project's root folder or if they are to be separated into different folders (`src` and `bin` by default). The latter is the default option.

You can create your project inside a working set. This is a good idea if you have too many projects in your workspace and want to keep them organized. Check the *Creating working sets* section of this chapter for more information on how to use and manage working sets.

The next page of the wizard contains build path options. In the *Managing the project build path* section of this chapter, we will talk more about these options. You can leave everything as the default for now, and make the necessary changes after the project is created.

Creating a Java class

To create a new Java class, right-click on the project in the Package Explorer view and navigate to **New | Class**. You will be presented with the **New Java Class** window, where you will input information about your class. You can change the class's superclass, and add interfaces that it implements, as well as add stubs for abstract methods inherited from interfaces and abstract superclasses, add constructors from superclasses, and add the main method.

To create your class inside a package, simply enter its name in the appropriate field, or click on the **Browse** button beside it and select the package. If you input a package name that doesn't exist, Eclipse will create it for you. New packages can also be created by right-clicking on the project in the Package Explorer and navigating to **New | Package**. Right-clicking on a package instead of a project in the Project Explorer and navigating to **New | Class** will cause the class to be created inside that package.

Creating working sets

Working sets provide a way to organize your workspace's projects into subsets. When you have too many projects in your workspace, it gets hard to find the project you're looking for in the **Package Explorer** view. Projects you are not currently working on, for example, can be kept in separate working sets. They won't get in the way of your current work but will be there in case you need them.

To create a new working set, open the Package Explorer's view menu (white triangle in the top-right corner of the view), and choose **Select Working Set**. Click on **New** and select the type of projects that the working set will contain (**Java**, in this case). On the next page, insert the name of the working set, and choose which projects it will contain. Once the working set is created, choose the **Selected Working Sets** option, and mark your working set. Click on **OK**, and the Package Explorer will only display the projects inside the working set you've just created.

Once your working sets are created, they are listed in the Package Explorer's view menu. Selecting one of them will make it the only working set visible in the Package Explorer. To view more than one working set at once, choose the **Select Working Set** option and mark the ones you want to show. To view the whole workspace again, choose **Deselect Workspace** in the view menu. You can also view all the working sets with their nested projects by selecting working sets as the top-level element of the **Package Explorer** view. To do this, navigate to **Top Level Elements | Working Sets** in the view menu.

Although you don't see projects that belong to other working sets when a working set is selected, they are still loaded in your workspace, and therefore utilize resources of your machine. To avoid wasting these resources, you can close unrelated projects by right-clicking on them and selecting **Close Project**. You can select all the projects in a working set by using the *Ctrl* + *A* keyboard shortcut.

If you have a big number of projects, but you never work with all of them at the same time (personal/business projects, different clients' projects, and so on), you can also create a specific workspace for each project set. To create a new workspace, navigate to **File | Switch Workspace | Other** in the menu, enter the folder name of your new workspace, and click on **OK**. You can choose to copy the current workspace's layout and working sets in the **Copy Settings** section.

Importing a Java project

If you are going to work on an existing project, there are a number of different ways you can import it into Eclipse, depending on how you have obtained the project's source code. To open the Import wizard, navigate to **File | Import**. Let's go through the options under the **General** category:

- **Archive file**: Select this option if the project you are working on already exists in your workspace, and you just want to import an archive file containing new resources to it. The Import wizard will list all the resources inside the archive file so that you can select the ones you wish to import. To select to which project the resources will be imported click on the **Browse** button. You can also select in which folder the resources are to be included. Click on **Finish** when you are done. The imported resources will be decompressed and copied into the project's folder.
- **Existing Projects into Workspace**: If you want to import a new project, select this option from the Import wizard. If the project's source file has been compressed into an archive file (the `.zip`, `.tar`, `.jar`, or `.tgz` format), there's no need to decompress it; just mark the **Select archive file** option on the following page of the wizard, and point to the archive file. If you have already decompressed the code, mark **Select root directory** and point to the project. The wizard will list all the Eclipse projects found in the folder or archive file. Select the ones you wish to import and click on **Finish**. You can add the imported projects to a specific working set and choose whether the projects are to be copied into your workspace folder or not. It's highly recommended to do so for both simplicity and portability; you know where all your Eclipse projects are, and it's easy to backup or move all of them to a different machine.

- **File System**: Use this wizard if you already have a project in your workspace and want to add new existing resources in your filesystem. On the next page, select the resources you wish to import by checking them. Click on the **Browse** button to select the project and the folder where the resources will be imported. When you click on the **Finish** button, the resources will be copied to the project's folder inside your workspace.
- **Preferences**: You can import Eclipse preferences files to your workspace by selecting this option. Preferences file contains code style and compiler preferences, the list of installed JREs, and the **Problems** view configurations. You can choose which of these preferences you wish to import from the selected configuration file.

Importing a project from Version Control Servers

Projects that are stored in Version Control Servers can be imported directly into Eclipse. There's a number of version control softwares, each with its pros and cons, and most of them are supported by Eclipse via plugins. GIT is one of the most used softwares for version control; we will discuss it with more details on how to use it efficiently inside Eclipse in *Chapter 4, Managing a Project*. **CVS** is the only version control system supported by default. To import a project managed by it, navigate to **CVS | Projects** from CVS in the Import wizard. Fill in the server information on the following page, and click on **Finish**.

Introducing Java views

As discussed in *Chapter 1, Introduction*, Eclipse's user interface consists of elements called views. The following sections will introduce the main views related to Java development.

The Package Explorer view

The **Package Explorer** view is the default view used to display a project's contents. As the name implies, it uses the package hierarchy of the project to display its classes, regardless of the actual file hierarchy. This view also displays the project's build path.

The following screenshot shows how the **Package Explorer** view looks:

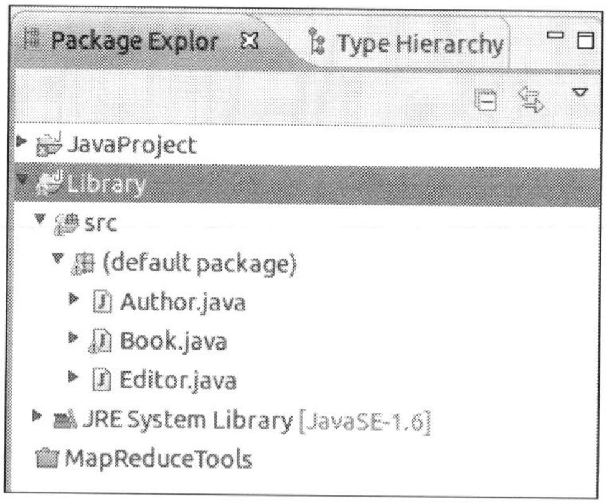

The Java Editor view

The Java Editor is the Eclipse component that will be used to edit Java source files. It is the main view in the Java perspective and is located in the middle of the screen.

The following screenshot shows the Java Editor view:

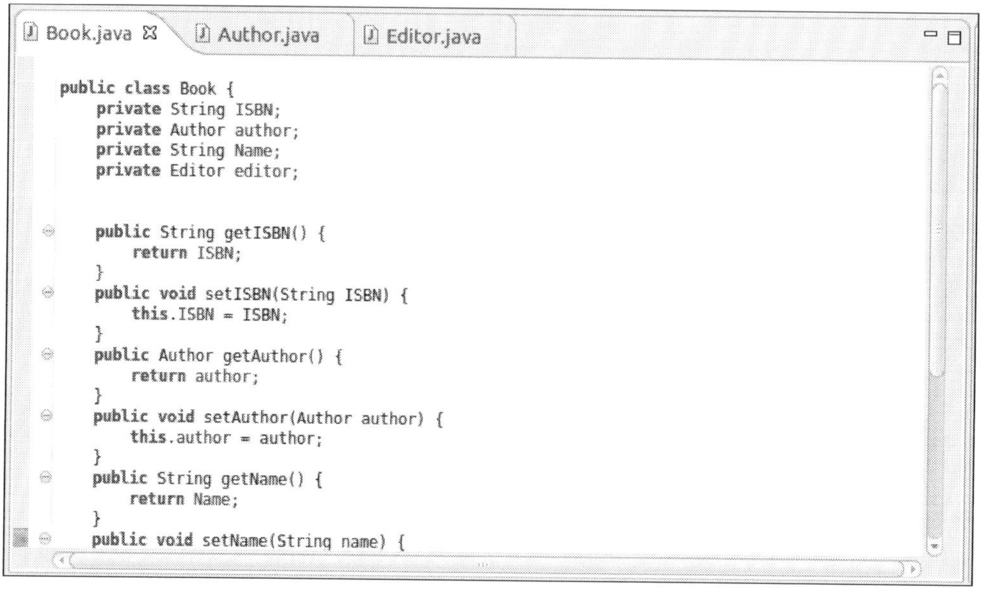

The Java Editor is much more than an ordinary text editor. It contains a number of features that makes it easy for newcomers to start writing Java code and increases the productivity of experienced Java programmers. Let's talk about some of these features.

Compiling errors and warnings annotations

As you will see in the *Building and running* section with more details, Eclipse builds your code automatically after every saved modification by default. This allows Eclipse to get the Java Compiler output and mark errors and warnings through the code, making it easier to spot and correct them. Warnings are underlined in yellow and errors in red.

Content assist

This is probably the most used Java Editor feature both by novice and experienced Java programmers. It allows you to list all the methods callable by a given instance, along with their documentation. This feature will work by default for all Java classes and for the ones in your workspace. To enable it for external libraries, you will have to configure the build path for your project. We'll talk more about build paths further in this chapter in the *Managing the project build path* section.

To see this feature in action, open a Java Editor, and create a new `String` instance.

```
String s = new String();
```

Now add a reference to this `String` instance, followed by a dot, and press *Ctrl* + *Space bar*. You will see a list of all the `String()` and `Object()` methods. This is way more practical than searching for the class's API in the Java documentation or memorizing it.

The following screenshot shows the content assist feature in action:

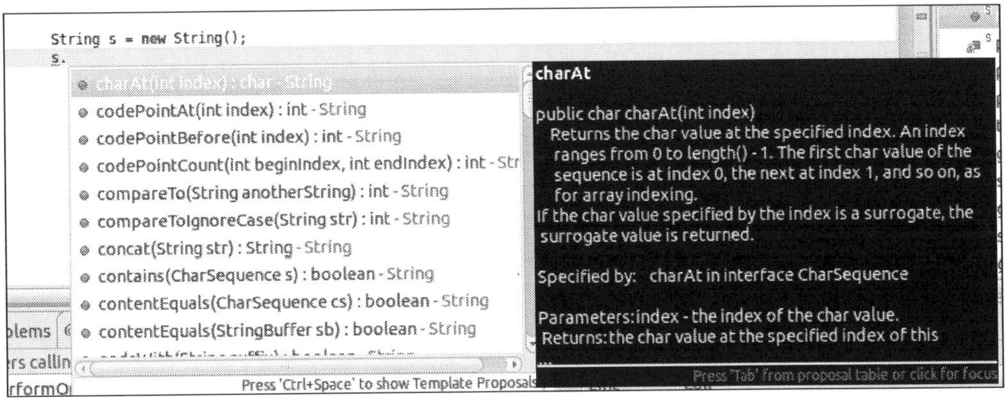

This list can be filtered by typing the beginning of the method's name after the dot. Let's suppose you want to replace some characters in this `String` instance. As a novice Java programmer, you are not sure if there's a method for that; and if there is, you are not sure which parameters it receives. It's a fair guess that the method's name probably starts with replace, right? So go ahead and type:

```
s.replace
```

When you press *Ctrl* along with the space bar, you will get a list of all the `String()` methods whose name starts with replace. By choosing one of them and pressing *Enter*, the editor completes the code with the rest of the method's name and its parameters. It will even suggest some variables in your code that you might want to use as parameters, as shown in the following screenshot:

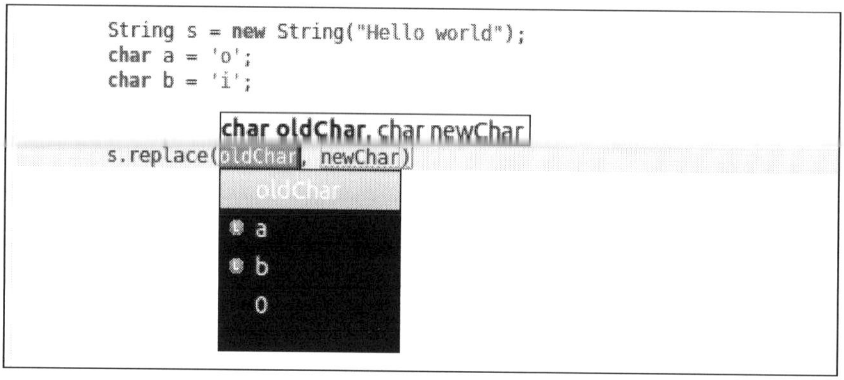

Content assist will work with all classes in the project's classpath. You can disable content assist's automatic activation by unmarking **Enable auto activation** inside the **Preferences** window and navigating to **Java** | **Editor** | **Content Assist**.

Code navigation

When the project you are working on is big enough, finding a class in the Package Explorer can be a pain. You will frequently find yourself asking, "In which package is that class again?". You can leave the source code of the classes you are working on open in different tabs, but soon enough you will have more open tabs than you would like to have.

Eclipse has an easy solution for this. In the toolbar, select **Navigate | Open Type**. Now, just type in the class's name, and click on **OK**. If you don't remember the full name of the class, you can use the wildcard characters, `?` (matches one character) and `*` (matches any number of characters). You can also use only the uppercase letters for the CamelCase names (for example, `SIOOBE` for `StringIndexOutOfBoundsException`). The shortcut for the **Open Type** dialog is *Ctrl* + *Shift* + *T*. There's also an equivalent feature for finding and opening resources other than Java classes, such as HTML files, images, and plain text files. The shortcut for the **Open Resource** dialog is *Ctrl* + *Shift* + *R*.

You can also navigate to a class' source file by holding *Ctrl* and clicking on a reference to that class in the code. To navigate to a method's implementation or definition directly, hold *Ctrl* and click on the method's call.

Another useful feature that makes it easy to browse through your project's source files is the **Link With Editor** feature in the **Package Explorer** view, as shown in the following screenshot:

By enabling it, the selected resource in the Package Explorer will always be the one that's open in the editor. Using this feature together with OpenType is certainly the easiest way of finding a resource in the Package Explorer.

Quick fix

Whenever there's an error or warning marker in your code, Eclipse might have some suggestions on how to get rid of it. To open the **Quick Fix** menu containing the suggestions, place the caret on the marked piece of code related to the error or warning, right-click on it, and choose **Quick Fix**. You can also use the shortcut by pressing *Ctrl* + *1* with the caret placed on the marked piece of code.

The following screenshot shows the quick fix feature suggesting you to either get rid of the unused variable, create getters and setters for it, or add a `SuppressWarnings` annotation:

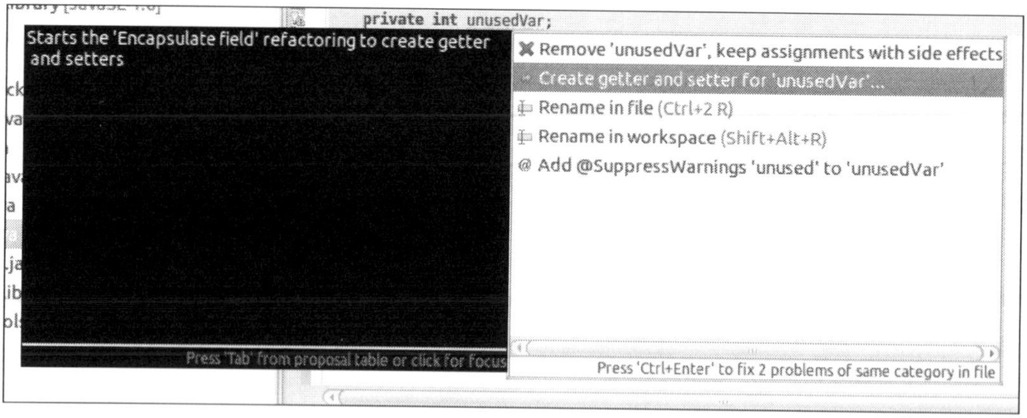

Let's see some of the most used quick fixes provided by Eclipse. You can take advantage of these quick fixes to speed up your code writing. You can for example, deliberately call a method that throws an exception without the try/catch block, and use the quick fix to generate it instead of writing the try/catch block yourself.

- **Unhandled exceptions**: When a method that throws an exception is called, and the exception is not caught or thrown, Eclipse will mark the call with an error. You can use the quick fix feature to surround the code with a proper try/catch block automatically. Just open the **Quick Fix** menu, and choose **Surround with Try/Catch**. It will generate a `catch` block that will then call the `printStackTrace()` method of the thrown exception. If the method is already inside a `try` block, you can also choose the **Add catch clause to the surrounding try** option. If the exception shouldn't be handled in the current method, you can also use the **Add throws declaration** quick fix.
- **References to nonexisting methods and variables**: Eclipse can create a stub for methods referenced through the code that doesn't exist with quick fix. To illustrate this feature's usefulness, let's suppose you are working on a class's code, and you realize that you will need a method that performs some specific operation with two integers, returning another integer value. You can simply use the method, pretending that it exists:

```
int b = 4;
int c = 5;
int a = performOperation(b,c);
```

The method call will be marked with an error that says `performOperation is undefined`. To create a stub for this method, place the caret over the method's name, open the **Quick Fix** menu, and choose **create method performOperation(int, int)**. A private method will be created with the correct parameters and return type as well as a TODO marker inside it, reminding you that you have to implement the method. You can also use a quick fix to create methods in other classes. Using the same previous example, you can create the `performOperation()` method in a different class, such as the following:

```
OperationPerformer op = new OperationPerformer();
int a = op.performOperation(b,c);
```

Speaking of classes, quick fix can also create one if you add a call to a non-existing class constructor.

Non-existing variables can also be created with quick fix. Like with the method creation, just refer to a variable that still doesn't exist, place the caret over it, and open the **Quick Fix** menu. You can create the variable either as a local variable, a field, or a parameter.

- **Remove dead code**: Unused methods, constructors and fields with private visibility are all marked with warnings. While the quick fix provided for unused methods and constructors is the most evident one (remove the dead code), it's also possible to generate getters and setters for unused private fields with a quick fix.

Customizing the editor

Like almost everything in Eclipse, you can customize the Java Editor's appearance and behavior. There are plenty of configurations in the **Preferences** window (**Window | Preferences**) that will certainly allow you to tailor the editor to suit your needs. Appearance-related configurations are mostly found in **General | Appearance | Colors and Fonts** and behavior and feature configurations are mostly under **General | Editors | Text Editors**. Since there are lots of different categories and configurations, the filter text in the **Preferences** window might help you find what you want. A short list of the preferences you will most likely want to change is as follows:

- **Colors and fonts**: Navigate to **General | Appearance**. In the **Colors and Fonts** configuration screen, you can see that options are organized by categories. The ones inside the **Basic** and **Java** categories will affect the Java Editor.
- **Enable/Disable spell checking**: The Eclipse editor comes with a spellchecker. While in some cases it can be useful, in many others you won't find much use for it. To disable or configure it, navigate to **General | Editors | Text Editors | Spelling**.

- **Annotations**: You can edit the way annotations (warnings and errors, among others) are shown in the editor by navigating to **General | Editors | Text Editors | Annotations** inside the **Preferences** window. You can change colors, the way annotations are highlighted in the code (underline, squiggly line, box, among others), and whether they are shown in the vertical bar before the code.
- **Show Line Numbers**: To show line numbers on the left-hand side of the editor, mark the corresponding checkbox by navigating to **General | Editors | Text Editors**. Right-clicking on the bar on the editor's left-hand side brings a dialog in which you can also enable/disable line numbers.

The Problems view

As we have seen in the Java Editor description, Eclipse uses the output of the automatic compilation of the class to mark the compiler's error and warning messages throughout the code. These messages are also listed in the **Problems** view. The following screenshot shows how the view looks:

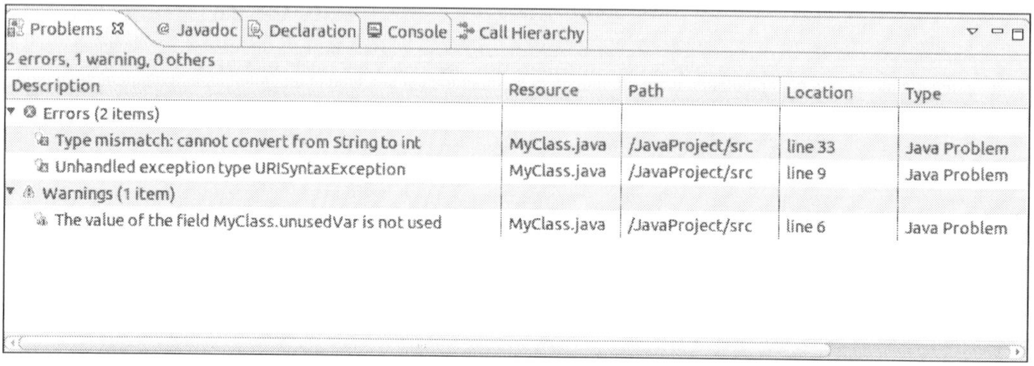

By default, it shows all the errors and warnings in all the open projects inside the current workspace. As you can see, every entry contains the compiler error/warning message and where the problem is located. By double-clicking on the entry, a new tab of the Java Editor, for the resource that contains the problem, will be opened (unless the tab already exists), and the caret will be placed on the problem's line.

Customizing the Problems view

The **Problems** view allows plenty of customization to make it easier to browse its entries. You can, for example, group them by severity and type by entering the **View** menu (white triangle in the top-right side of the view) and selecting the **Group By** entry.

When you have too many different projects open in your workspace, showing all the errors and warnings from all the projects might not be a good idea. Fortunately, you can customize what the **Problems** view will show. For this, open the **Configure Contents** window in the view menu.

There's a list of problem configurations on the left-hand side of the window. You can choose to show items that match either all or any of the configurations. New configurations can be added by clicking on the **New** button. The four default configurations cannot be removed or renamed, but they can be edited just as the ones you created.

You can modify the scope in which your configuration will search for problems and the type of problem that your configuration contains (some of these types may not refer to Java code). You can also filter problems that either contain or do not contain some text.

Another feature of the **Problems** view that helps organize the problems' listing is creating other views. To create another view, open the view menu and choose the **New Problems** view. Each view will have its own configuration, so you can create multiple views and have each one displaying one type of problem.

The Outline view

The **Outline** view provides an overview of the selected class. By default, it shows a list of the class's fields and methods. You can configure the view to hide static elements, fields, nonpublic members, and local types.

The following screenshot shows the outline of a `Book` class:

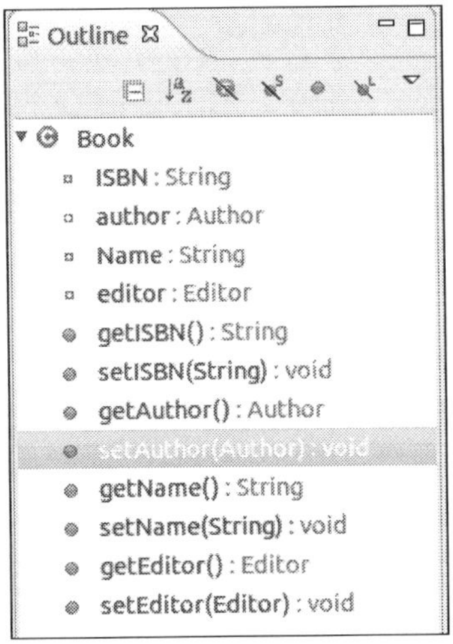

The Type Hierarchy view

The **Type Hierarchy** view is where you visualize a class's superclasses and extensions. To load a class's type hierarchy in this view, select the reference to the class name in the Java Editor, right-click on it, and select **Open Type Hierarchy**; alternatively, you can use the *F4* keyboard shortcut. This tool can be useful when you have an interface and want to list all of its implementations.

The following screenshot contains the **Type Hierarchy** view showing the supertype hierarchy of the `java.awt.Button` class:

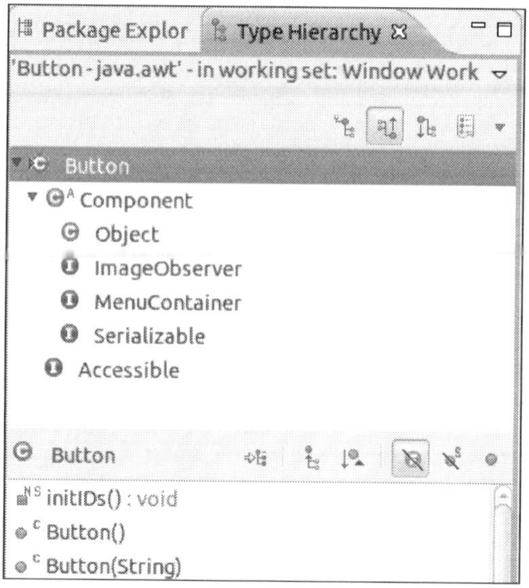

There's an alternative to the **Type Hierarchy** view named **Quick Type Hierarchy**. Instead of loading and displaying the **Type Hierarchy** view, it shows the class's hierarchy in a pop-up window. The keyboard shortcut for **Quick Type Hierarchy** is *Ctrl* + *T*.

The Call Hierarchy view

To see all the calls made to a specific method in the project, select a reference to the method's name in the Java Editor, right-click on it, and select **Open Call Hierarchy**; alternatively, use the *Ctrl* + *Alt* + *H* keyboard shortcut. The view provides a full, recursive hierarchy of the selected method's calls, as shown in the following screenshot:

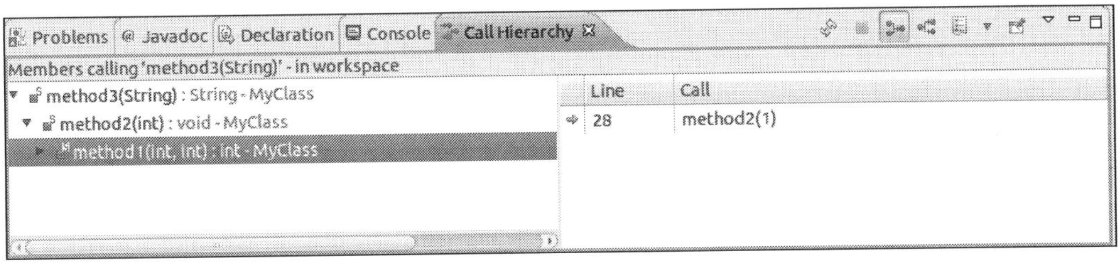

Organizing imports

Eclipse makes it easy to manage a class's imports. By right-clicking on the Java Editor and navigating to **Source** | **Organize Imports**, all unused imports and import classes are removed if there are references to them in the code. You can also use the *Ctrl* + *Shift* + *O* shortcut to organize the imports.

To illustrate how this functionality is useful, let's suppose you want to add an `ArrayList` instance in your code. Instead of importing `ArrayList` and using it (and you may or may not know the package name of `ArrayList` from the top of your head), you can simply write:

```
ArrayList a;
```

Alternatively, you can also press *Ctrl* + *Shift* + *O* to add the import.

If the class's name occurs in more than one package in the current classpath, you will be prompted to choose the one to which you're referring.

Save actions

You can define actions that are to be performed every time a Java source file is saved. For this, navigate to **Window** | **Preferences** in the toolbar; then, in the left menu, navigate to **Java** | **Editor**, and select **Save Actions**. Check the **Perform the selected actions on save** checkbox, and choose the actions you want to perform. It's possible to organize imports automatically to remove unused ones, add missing annotations, format the source code, and reinforce code style.

You can also define specific save actions for each project by right-clicking on the project's entry in the Package Explorer and clicking on **Properties**. Then, on the new window, navigate to **Java Editor** | **Save Actions**, and check **Enable project specific settings**.

Enforcing Coding Style with Formatter

Some companies and open source projects are very strict with their coding style. They have clear guidelines on when to skip lines or not, where to put whitespaces, how the indentation should look, and so on. All these guidelines can be implemented in Eclipse's Formatter tool that helps the developers to style the code without any worry. The formatter will modify your code to comply with a set of coding style rules. To run the formatter, navigate to **Source | Format** in the toolbar or use the *Ctrl + Shift + F* shortcut. You can also force Eclipse to run the formatter after file saves using **Save Actions**.

To edit your workspace's formatting settings, go to the preferences window (**Window | Preferences**), and navigate to **Java | Code Style | Formatter** in the left menu. The formatter's settings are saved under profiles that can be saved, imported, and exported. The **Edit** button brings up a huge number of settings that will suit most of your code-styling needs.

You can also define project-specific format rules. In the project properties window (right-click on the project in the Package Explorer and select **Properties**), navigate to **Java Code Style | Formatter** in the left menu, and check **Enable project specific settings**.

Generating code

In Java, you will eventually find yourself writing similar pieces of code over and over. Even though the good usage of design principles tends to minimize this, there are some methods, such as getters, setters, and constructors, in most cases look exactly the same. Eclipse can help you by generating code for these simple methods.

You can see all the code that Eclipse can generate by right-clicking anywhere in the Java Editor and choosing the **Source** entry. This menu can also be opened with the shortcut *Alt + Shift + S*.

Generating getters and setters

Getters and setters are generally the same. They provide access to some private field of the class by either returning its value or by assigning a new one. To avoid this no-brainer task and save some time to write real code, choose **Generate Getters and Setters** from the **Source** entry. You will be presented with a window, as shown in the following screenshot:

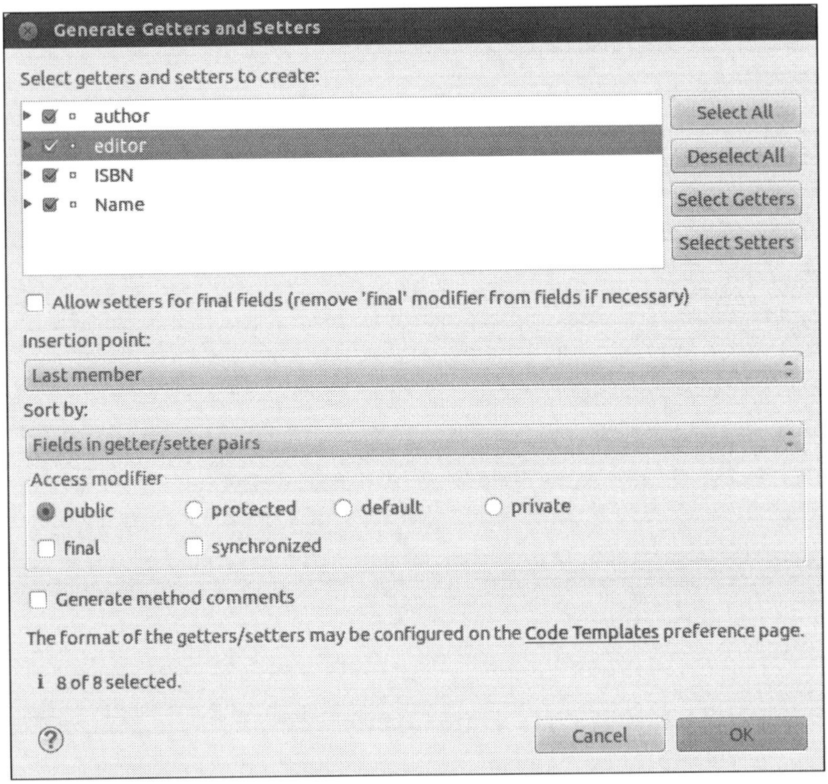

In the **Generate Getters and Setters** window, you will be able to determine for which fields you want to generate getters and setters, as well as choosing their access modifiers and where in the code they will be inserted.

Generating constructors

Just like getters and setters, class constructors have a very typical template that receives values to be assigned to the class's fields and calls the superclass's constructor. You can also skip this tedious task by clicking on **Generate Constructor using Fields** in the **Source** menu entry. You will be presented with a window similar to the **Generate Getters and Setters** one in which you will be prompted to choose which fields will have their values initialized by the constructor, and where the constructor will be inserted in the code, among others.

Generating the hashCode() and equals() methods

There are cases in which you have to override the `equals()` and `hashCode()` methods inherited from the `Object` class in order to treat two different instances as being equal when they have the same parameter values.

Eclipse also has a template for these two methods. By choosing **Generate hashCode() and equals()** inside the **Source** menu entry, you will be able to generate methods that utilize the fields of your choice to compare two instances and to generate the hash code of the instance. The template for the `equals()` method will check if the other object is null, if it is an instance of the same class, and if the selected parameters are equal.

Generating the toString() method

To generate a `toString()` method that returns the value of the class's fields in a human-readable form, choose the **Generate toString()** option. Like the previous generators, this one will allow you to choose which fields you want your `toString()` method to contain.

Generating method comments

Comments for methods in the `Javadoc` format can easily be generated in Eclipse. For this, right-click on the desired method, and navigate to **Source** | **Generate Element Comment,** or use the shortcut *Alt* + *Shift* + *J* with the caret positioned in the method's signature line. These comments will be shown in Eclipse when the user hovers the mouse over the method's call or when the content assist feature is used.

The generated comment contains the method's parameters and return type. You only have to add a simple description of the method's usage and some words about the parameters when required. An example of a generated comment is presented in the following screenshot:

> These shortcuts might look useless and hard to memorize, but once you get used to them, they become incredibly handy and increase your coding speed considerably since you'll rarely have to take your hands off your keyboard when you know the ones you use most by heart. The *Appendix, Keyboard Shortcuts* contains an extensive list of all the shortcuts presented in this book.

```
/**
 * @param isbn
 * @param author
 * @param name
 * @param editor
 */
public Book(String isbn, Author author, String name, Editor editor) {
    super();
    this.isbn = isbn;
    this.author = author;
    this.name = name;
    this.editor = editor;
}
```

Editing code and comment templates

Although these templates will work on most of the situations, your project might have peculiarities that require a specific type of constructor. You can still take advantage of code and comment generation by editing the templates and tailoring them to fit your needs. For this, right-click on the project's entry in the Package Explorer and choose **Properties**. Expand the **Java Code Style** entry in the left menu and choose **Code Templates**. The following window will open:

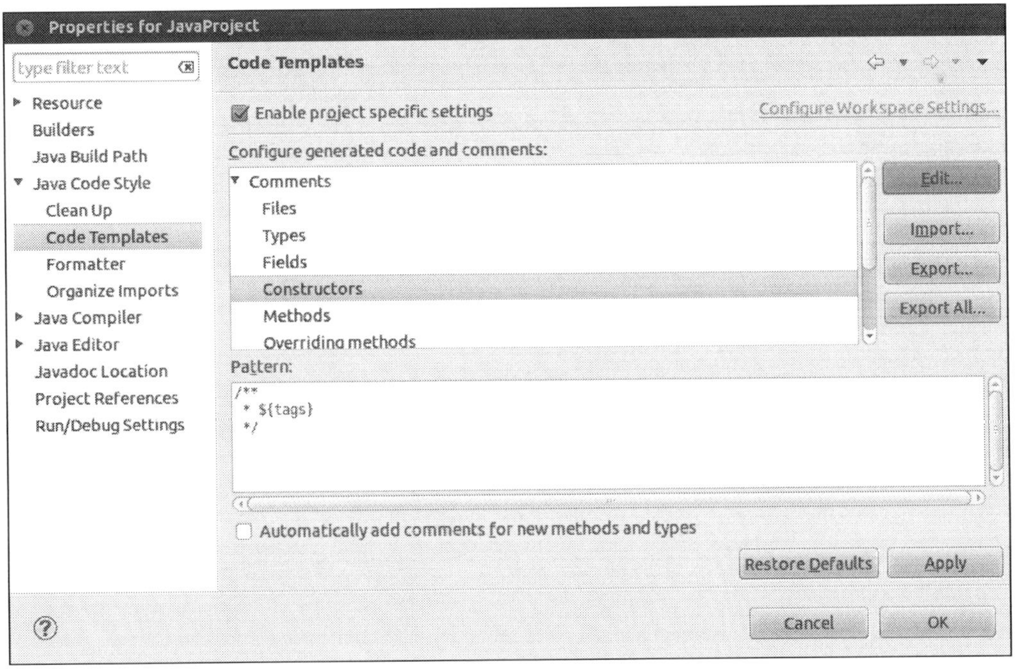

By choosing a type of generated code or comment and clicking on **Edit**, you will be able to add code and variables to the template. You can also choose to add comments on newly generated methods and types automatically.

The code template settings edited in this window are project specific. The settings are saved in a metadata file inside the project folder, so when a project is imported, the coding style settings are automatically set. This allows you to enforce the project's coding style among all the project's developers. If you want to edit the default code templates for the whole workspace, open the **Preferences** window and navigate to **Java | Code Style | Code Templates**. Projects that don't contain any specific setting for templates will use these settings.

Refactoring

Code refactoring can be a real pain. As your project grows, adding a parameter to a method or renaming it can turn into a nightmare. Eclipse has some refactoring tools that make such tasks a walk in the park.

Renaming variables

Renaming variables manually can be tedious and time consuming. You could resort to find/replace to make it a little less painful. But since the variable name might be matched in other places of the code, you have to go through each match of the variable name and verify if it has to be modified (try renaming a variable named `x` in a big enough class and you'll understand). If the variable is public, you would also have to scan the whole project for references of this variable.

Luckily, Eclipse does this for you. Just right-click on any occurrence of the variable in the code and navigate to **Refactor | Rename**. Type the new variable's name and hit *Enter*.

You can see that even if the name's variable occurs in other parts of the code, only the text that's within the variable's name is modified. Eclipse is only able to do that because it compiles the code and uses information from this process to know which ones are references of the variable in the code. Because it relies on the compiled code, it's not possible to rename a variable inside a class with compiling errors. Eclipse also won't allow you to change a variable's name to one that already exists in the same context. Renaming also works with other elements, such as classes, methods, and interfaces.

> Don't let Eclipse's features soften you! Just because it's super easy to rename a variable, you shouldn't feel free to ignore good code practice, such as variable encapsulation. Good code is good code, regardless of the IDE.

Modifying a method's signature

Modifying a method's signature poses similar inconveniences for renaming variables. Eclipse can also help in that task. Right-click on the method's name, and navigate to **Refactor** | **Change Method's Signature**. You will be presented with a window that allows you to change the method's access modifier, return type, method name, and add/remove parameters.

On modifying a method's signature, all calls to that method are also modified. When you add a new parameter to the signature, you can choose a default value that will be added to the calls.

Building and running the project

By default, Eclipse builds your project's code automatically after every saved modification. It uses a customized internal compiler with some interesting features, such as executing a class even when it contains errors and incremental compilation (which allows compiling just parts of the project). After the modifications have been recompiled, it can highlight the problems with the code on the fly. You can see a list of these errors in the **Problems** view. If the view is not currently visible in the workspace, you can open it by navigating to **Window** | **Show View** | **Problems**.

However, if you are working on a project that's too big, or if you are using a machine with low resources, automatic building can slow things down. To disable it, click on **Project** and unmark the **Build Automatically** option. Once you do this, the project will only be built when you navigate to **Project** | **Build Project** or **Project** | **Build All**.

To run your project, select it in the **Package Explorer** view, and click on the green play button in the toolbar, or select **Run** from the **Run** tab. This will launch your project without any further configuration. If your project contains a class with a `public static void main(String[] args)` method, it will be executed. If more than one class contains it, you will be able to choose which one you wish to run. You can see the output of your program (if there is any) in the **Console** tab.

Creating a launch configuration

If your project needs customizations, such as arguments or a specific classpath configuration to run, you can create a launch configuration to meet these needs. For this, navigate to **Run | Run Configurations**. The following screenshot shows the launch configuration window:

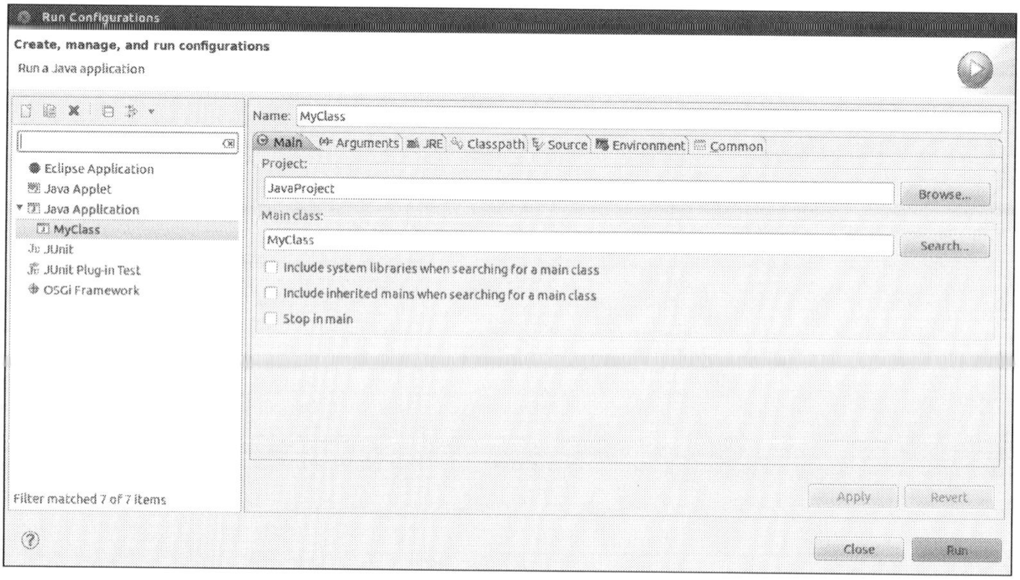

On the left-hand side of the **Run Configurations** window, you have a list of all the launch configuration categories. To create a new launch configuration for a Java project, double-click on the **Java Application** category. If you have already run your application once with the **Run** button, there will be a default configuration in this category.

By selecting the configuration, you will be presented with the tabs that contain the launch configurations. A brief description of what each tab contains is as follows:

- **Main**: This is where you will choose the project and the main class. The **Search** button beside the **Main class** text field allows browsing through all the classes with a main method within the project.
- **Arguments**: This tab lets you set arguments for both the main class itself and for the JVM.
- **JRE**: This tab lets you choose the JRE that will be used to run the class.
- **Classpath**: Edit this tab to create a launch configuration with a specific classpath configuration. Modifying this tab's configurations doesn't modify the project's classpath configuration.

- **Source**: This tab lets you change the source lookup path. As in the **Classpath** tab, these configurations don't modify the project itself and are only valid for this launch configuration.
- **Environment**: This tab allows you to define environment variables for this run configuration.
- **Common**: This tab contains the general configuration of the launch. It allows you to choose to allocate a console for the launch and write the output to a file instead of the console. The **Save as** section refers to the launch configuration file.

Managing the project build path

The build path is a list of JARs, classes, and source file folders in which Eclipse will look to resolve dependencies. You can edit the build path for a project by entering the project's properties and selecting **Java Build Path**. The configuration options have been divided into four tabs:

- **Source**: This contains options to manage the project's source folders. It allows you to create new source folders and add links to external sources in the filesystem.
- **Projects**: You can manage you project's dependency on other projects in your workspace in this tab. When you'll add a dependency to another project, you'll have full access to all of its classes by default. However, some of these classes might not be visible when the project is compiled and executed outside Eclipse. You might end up with a project that works inside Eclipse but doesn't work when you run the JAR files outside it. To avoid these head-scratching issues, you can create access rules to external projects to keep them from using other projects' internal packages. For this, expand the required project, choose **Access Rules**, and click on **Edit**. Click on **Add Access Rule** and add **Rule Pattern**. You can discourage (compiles but shows a warning) or forbid (compiling fails with an error) access.
- **Libraries**: In this tab, you'll be able to edit the libraries that will be added to the project's build path. You can add JARs and class folders from inside or outside the workspace, and you can also add libraries and variables. You can add a link from a library to its Javadoc. This will allow you to have methods and classes with comments that describe their usage, as if they were in your workspace. To add a Javadoc for a library, expand its entry in the list, select **Javadoc location**, and click on the **Edit** button. Enter the path for the Javadoc and click on **OK**. You can also point to the library's source if you have access to it.

- **Order and Export**: This is where you'll define which source folders and libraries from the project will be exported so that other projects can import them. This tab also allows you to choose the order in which the libraries will be loaded during runtime. The order will determine which class will be loaded in case two different libraries provide that same class. To increase or decrease a library's priority, select it and press the **Up** and **Down** buttons.

Summary

In this chapter, we have covered the basics of Java developing with Eclipse: creating and importing a Java project, organizing projects in working sets, adding new classes, among others. We went through all of Eclipse's views for Java developing, learned how they can be used to navigate through code, found errors, warnings, and much more. Special attention was given to the Java Editor by covering its most useful features. We also covered some more advanced topics, such as creating templates for code generation and project settings, that can be shared among developers in a team to enforce code style and best practices.

In the next chapter, you will learn how Eclipse can help you debug your Java applications as well as test them using JUnit.

3
Unit Testing with JUnit and Debugging

By now, you are past the basics and should be familiar with developing Java applications using the Eclipse IDE. Although you must be feeling very confident with your recent acquired skills, before we go any further, you should know that the more you code, the higher are the chances of you introducing a bug into your code. Even with all the facilities provided by Eclipse, things are bound to go wrong as Java applications scale up. So, we need to build a safety net whenever our application code starts to become nontrivial to manage. One way of doing this is to test our programs, and testing is one of the things that we are going to cover in this chapter.

In an ideal world, all problems detected by our tests would be easy to track down and fix. Nevertheless, this is not always the case. It turns out that nasty bugs might make their way into our code. As mentioned, having a good set of tests helps us to spot problems, however, sometimes the feedback provided by the test set is not enough, and we need to dig deeper to uncover problems. The act of understanding why the software is not behaving properly and working out a fix for it is called **debugging**. Debugging is covered later in this chapter.

When it comes to testing and debugging Java programs, Eclipse has got you covered. It comes with built-in features for automating the creation and execution of JUnit tests, and it features an impressive Java debugger.

We will cover the following topics in this chapter:

- Configuring JUnit inside Eclipse
- Organizing, developing, and running unit tests using JUnit inside Eclipse
- Using Eclipse's built-in debugger when bugs arise and understanding that only test methods are not enough to pinpoint the problem cause

Testing with JUnit – getting started

Before describing JUnit, we believe that some words about unit testing are in order. When carrying out unit testing activities, emphasis is given to small chunks of code. In other words, you are not trying to cover the whole program with a single test. Rather, you are testing the foundations of your programs by writing a suite of tests per class, one or more tests per method. Writing tests while you are working on the code at hand has a host of benefits. For instance, it is easier to find bugs when unit testing. This is the case because each unit test should evaluate exactly one functional unit, which makes the scope one has to hunt for bugs significantly smaller than searching the whole program for bugs. However, according to Robert C. Martin, to reap the full benefits of unit testing, your test suite has to comply with the following principles (also known as the **FIRST** acronym):

- **Fast**: The tests should be fully automatic and should be able to check their own results. But most importantly, you should be able to execute them fast so that they can be frequently executed.
- **Independent**: The tests should not depend on each other you should be able to run your tests in any particular order.
- **Repeatable**: You should be able to execute your tests in any environment (not only production) and as frequently as required.
- **Self-validating**: The tests should produce a Boolean outcome, that is, they should either pass or fail; there should be nothing in between. In fact, it boils down to avoiding subjective outputs that require manual evaluation.
- **Timely**: According to Robert C. Martin, unit tests should be written before the code implements the feature that they evaluate. According to him, it is harder to write tests after implementing the production code. From an academic perspective, there is no substantial evidence regarding the benefits of writing tests before, in comparison to writing tests after implementing the code that they evaluate. This is clearly a controversial topic, we do not seek to influence you in favor of a particular approach. Try writing tests before and after implementing the code and see which works best for you.

> In case you think that writing tests before implementing code suits you better, you should read about **Test Driven Development** (**TDD**). While you are at that, you might also want to study refactoring, which consists of improving your code without changing its external behavior.

Now that you have a basic understanding of the philosophy that backs up unit testing, you are ready to take the next step towards writing your own tests—learning how to use a unit testing framework. Since JUnit is arguably the most popular testing framework in the Java community, we will cover it here.

More specifically, JUnit is an open source testing framework, which was initially contrived by Kent Beck and Erich Gamma (`www.junit.org`). As of the time of this writing, the latest stable version of this framework (and the one bundled with the IDE) is Version 4.10. Due to the fact that Erich led the design of the Eclipse **Java Development Tools (JDT)**, JUnit is well-integrated into the IDE. Eclipse already ships with a bundled JUnit JAR. So, as long as you are planning on using JUnit 4.10, there is not much to say about installing it. However, it is also possible to use a different JUnit version. The steps for setting up the default and a non-default JUnit JAR are covered in the following sections.

Although the duo, JUnit and Eclipse, supports and automates most of the testing process, you still need to write your own test files manually. Since Eclipse leaves the organization of your test files up to you, a word about how to organize these files is in order.

JUnit-based test files can be organized in the following way. Suppose your application has a class named `MyAppClass`, and you want to test it. To do so, you need to place your tests in another class named `MyAppClassTest`. Usually, test classes are kept separated from regular code. The convention is placing them into a subpackage called test. So, let's say that `MyAppClass` is in a package called `org.myapp`, tests related to this class need to be placed in the subpackage named `org.myapp.test`. As for the tests themselves, they are grouped in `MyAppClassTest` in the form of Java methods with some slight modifications that we will describe shortly. An overview of this typical organization is shown in the following screenshot, which illustrates this organization using a hypothetical Java project named `MyTestProject`:

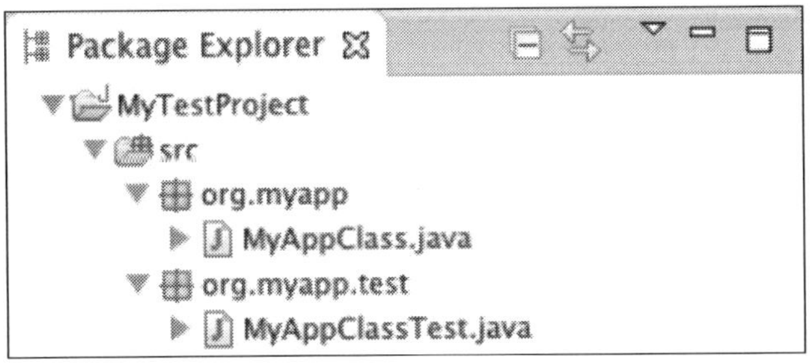

> It is worth mentioning that neither JUnit nor Eclipse enforces this naming convention. The creation of separate folders/subpackages for tests is not mandatory either. However, following these conventions is a good programming practice. It is also worth mentioning that another widely used convention to organize test code is the maven convention. By following this convention, you need to create separate folders for regular Java code and test code, `src/main/java` and `src/test/java`. Also, the test classes are placed in the same package in which the classes they test are placed.

Setting up JUnit

In order to use JUnit with Eclipse, you need to add `junit.jar` to your project's build path. To do this, right-click on your project, and navigate to **Build Path | Add Libraries...**, as shown in the following screenshot:

The following screenshot shows the options listed by Eclipse. Select the second option **JUnit**. You can select either JUnit Version 3 or Version 4.

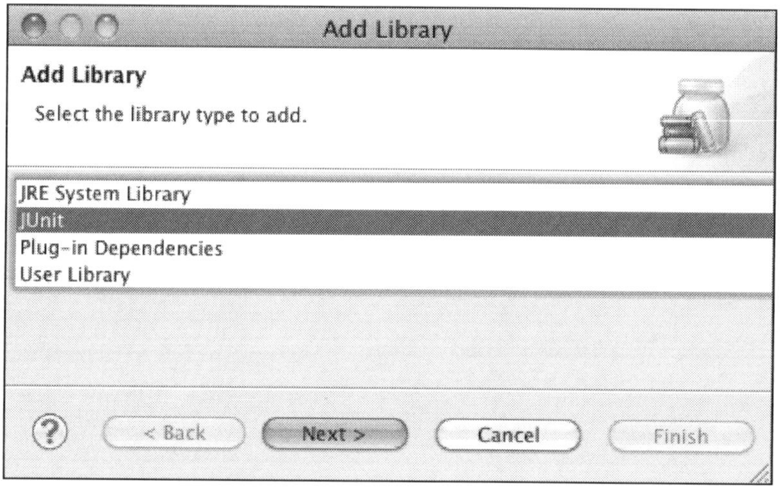

Throughout this chapter we are going to use JUnit 4 (4.10 to be more specific); so go ahead and select **JUnit 4** as shown in the following screenshot:

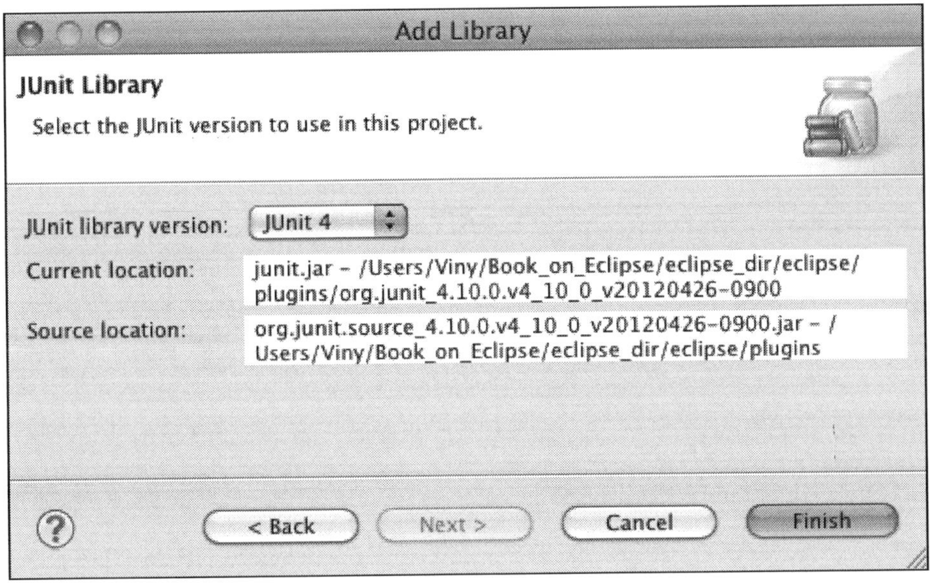

There is a striking difference between JUnit 3 and JUnit 4, mostly due to the fact that JUnit 4 heavily relies on the new features introduced by the most recent versions of Java. To be more specific, the main difference between JUnit 3 and JUnit 4 is that the latter relies on annotations rather than hierarchies and method-name conventions. If you are familiar with JUnit 3, but have not used JUnit 4 yet, you will notice that the latter is simpler and easier to use than the former.

By choosing JUnit 4, Eclipse adds JUnit 4.10 to your project's build path. If you need to use a different JUnit version, you can download that particular version from JUnit's website and add it to your project's build path. To do that, download the desired version from JUnit's website, right-click on your project name, and navigate to **Build Path | Configure Build Path...**, as illustrated in the following screenshot:

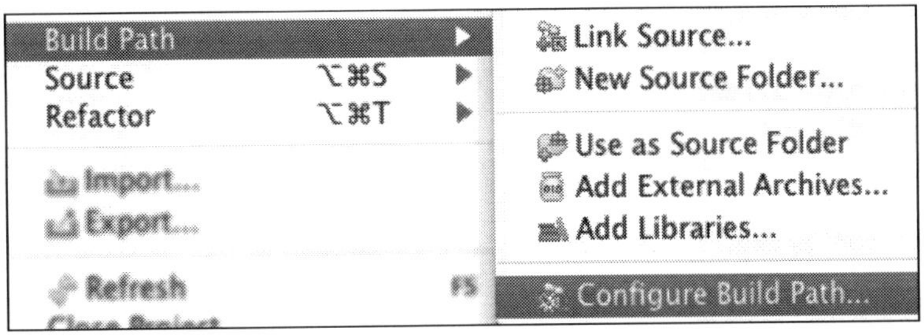

The project properties window will appear, showing the project's build path current configurations (shown in the following screenshot). Select the **Libraries** tab, click on **Add External JARs...**, locate and select your recently downloaded JAR, and then click on **OK**. Click on **OK** again to close the properties window. Done, now you have JUnit up and running. We are all set to create JUnit-based tests. It is worth mentioning that through the project properties window you can also add JUnit 4 by clicking on the **Add Library...** option. Selecting this option opens the pop-up window shown in the earlier screenshot.

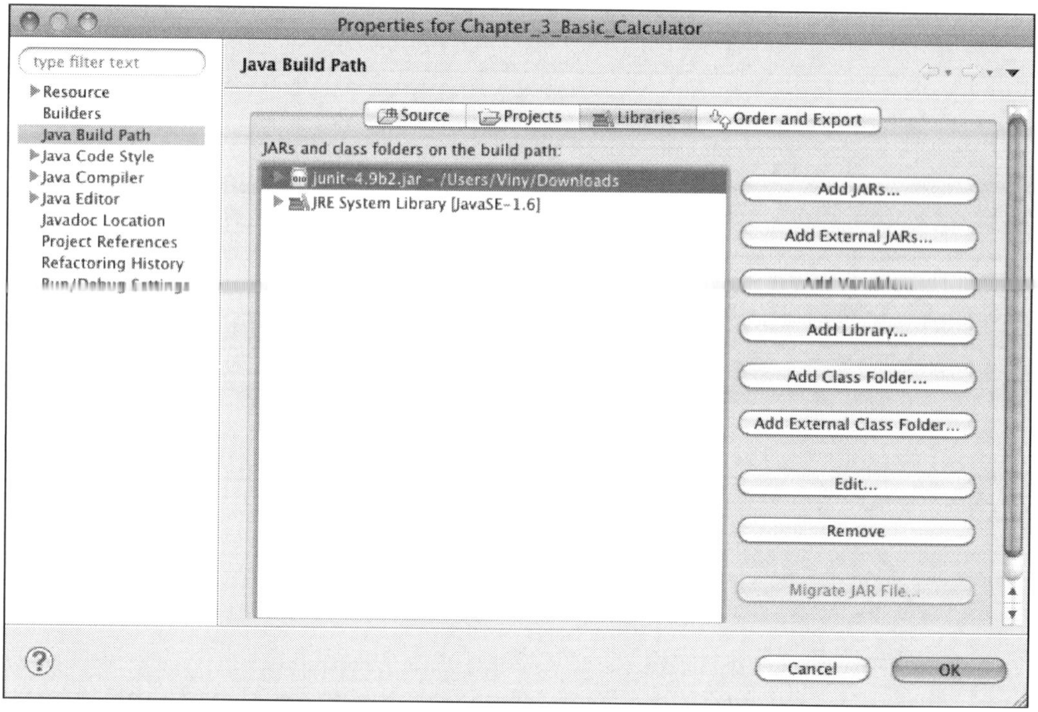

Adding a new JAR to the build path can be done in a myriad of ways. In this section we just covered two. If you followed the first approach, the JUnit JAR file is going to be placed under **JUnit 4** as shown in **(a)** in the following screenshot. Otherwise, the JAR file is going to appear under **Referenced Libraries** as shown in **(b)**. Keep on reading to discover an even easier approach for setting up JUnit.

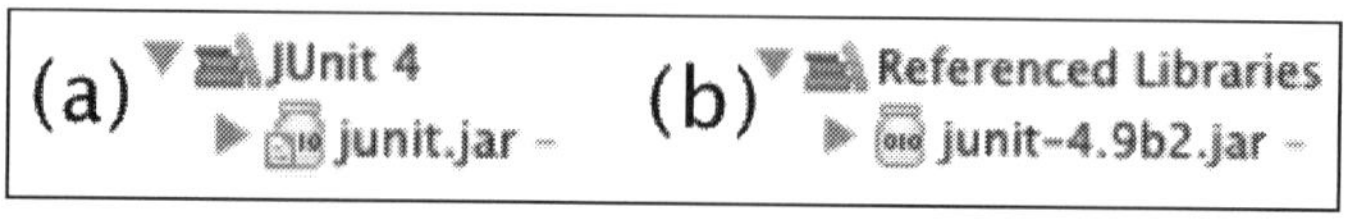

Testing with JUnit

Using JUnit, it is remarkably simple to write repeatable tests. Actually, from day one, being simple to use was a key design decision made by Kent and Erich. According to them, if JUnit was not easy to learn and execute, programmers would not actually use it.

Basically, when you need to test something using JUnit, you perform these basic actions:

1. Create the Java class that is going to enclose all your tests. Make sure to name it according to the convention we described in the previous section.
2. Create a method and annotate it with `@org.junit.Test`.
3. Write the code related to the behavior that you want to test inside the previously created method.
4. To check whether the code under test yielded the expected result, import `org.junit`. Assert statically, and invoke one of the methods of the Assert family (`assertTrue`, `assertEquals`, and so on) to compare the result of your test with the expected result.

The JUnit framework has several other features, but these are the ones you will need to understand in order to get started. So, let's dive into the technical details of these basic features by testing a Java class that implements a basic calculator.

First, we need a new Java project for our calculator class. So, inside Eclipse navigate to **File** | **New** | **Java Project**. To give our project a name, type `BasicCalculatorProject` in the **Project name** field, and then click on **Finish**. You might be prompted to switch to the Java Perspective. If this happens, answer **Yes**. After performing all these steps, the **Package Explorer** view should show your new Java project. However, if something goes wrong and you can run into problems, go back to the previous chapter and review how to create a Java project.

Inside our recently created Java project, create a new package. To create a package named `chapter3.basiccalc`, select the `src` folder, right-click on it, and navigate to **New** | **Package**. Enter the package name, and click on the **Finish** button.

Let's create a class named `BasicCalculator`. Right-click on your recently created package and navigate to **New** | **Class**. Our class under test is as follows:

```
package chapter3.basiccalc;

public class BasicCalculator {
  private static int[] tempVars; //acts as an array of temp
variables
  private static int result;

  public void add(int n) {
    result += n;
  }

  public void subtract(int n) {
    result -= n;
  }

  public void addResultToTempAt(int index) {
    tempVars[index] = result;// BUG: tempVars wasn't initialized
properly
  }

  public void squareRoot(int n) {
    //not implemented yet
  }

  public void divide(int n) {
    result /= n;
  }


  public void multiply(int n) {
    result = n * n;
  }

  public void clear() {
    result = 0;
  }

  public void clearTemps() {
    for (int i =0; i < tempVars.length; i++) {
      tempVars[i] = 0;
    }
  }
```

```
    public int getResult() {
      return result;
    }

  }
```

> **Downloading the example code**
>
> You can download the example code files for all Packt books you have purchased from your account at `http://www.PacktPub.com`. If you purchased this book elsewhere, you can visit `http://www.PacktPub.com/support` and register to have the files e-mailed directly to you.

Our calculator implementation contains two faults, but do not mind about them yet. Using a buggy implementation will make our example more interesting.

The purpose of the `BasicCalculator` class is to perform elementary arithmetic operations (addition, subtraction, multiplication, and division) and calculate the square root; while storing the result after each operation. Apart from the result variable, an array is used to store preliminary results; making it possible to reuse it in subsequent calculations. Methods for reusing the values stored in this array are not implemented yet. So far, our calculator can only store preliminary results (which are kept in the `tempVars` array) and clean its memory (by assigning zero to all elements of `tempVars`). Further, notice that the method that should calculate the square root is not implemented either.

> Well, our class should calculate the square root, however, as you saw, this particular feature has not been implemented yet. Do not worry about that, we did not implement this particular feature for demonstration purposes. Later in this chapter, we will show you how to make JUnit skip a test case for which either there is no implementation or the current implementation has been acting up.

Since we are sticking to the naming convention we presented in the previous section, before testing `BasicCalculator`, you should create a package called `chapter3.basiccalc.test`. The class containing all test cases should be created inside this package.

Now that you have set your Java project up and dealt with all scaffolding around your test code, you can either add JUnit to your project build path following the aforementioned directions, or keep reading and discover an even easier way of adding JUnit to the build path (as we mentioned, there are many ways of doing operations in Eclipse).

Right-click on `BasicCalculator` and navigate to **New | JUnit Test Case**. The window shown in the following screenshot will appear:

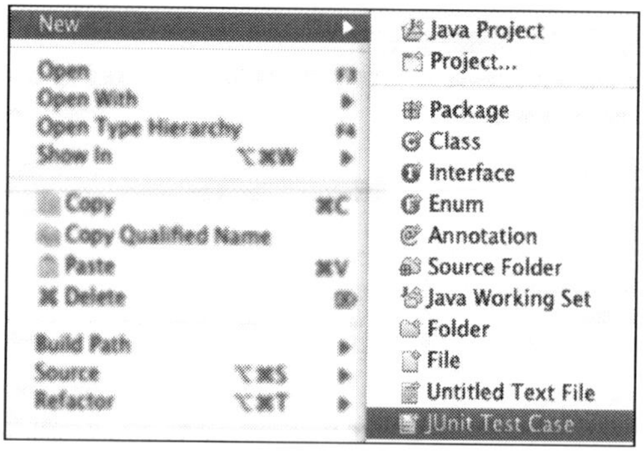

After selecting the **JUnit Test Case** option, a new window will pop up with the name of the test case already filled out (`BasicCalculatorTest`). It also shows the class under test (`BasicCalculator`). You only need to type the test package name in the **Package** field. So go ahead and type `chapter3.basiccalc.test` as we did in the following screenshot (or click on the **Browse** button and select the desired package).

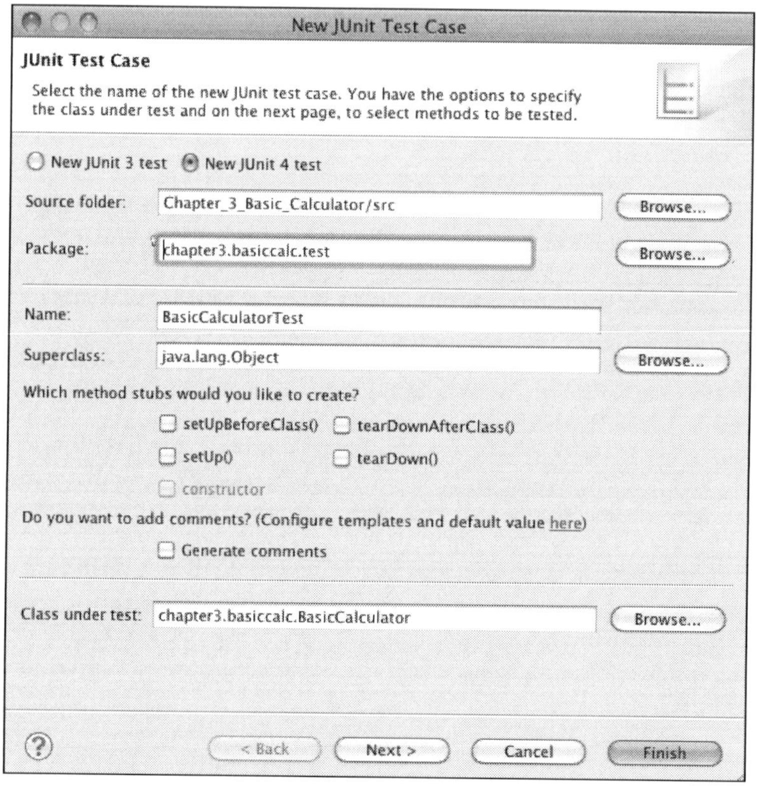

Implementing the test methods generated by Eclipse

Notice that we uncheck the **setUp()** and **tearDown()** options. We are going to explain what these terms mean shortly. But for the time being, we do not want Eclipse to automatically generate these methods for us. After unchecking these options (in case they are not unchecked by default), click on **Next** and select the methods that you want to test, as demonstrated in the following screenshot. Here Eclipse is prompting you to select the methods for which it should generate stubs.

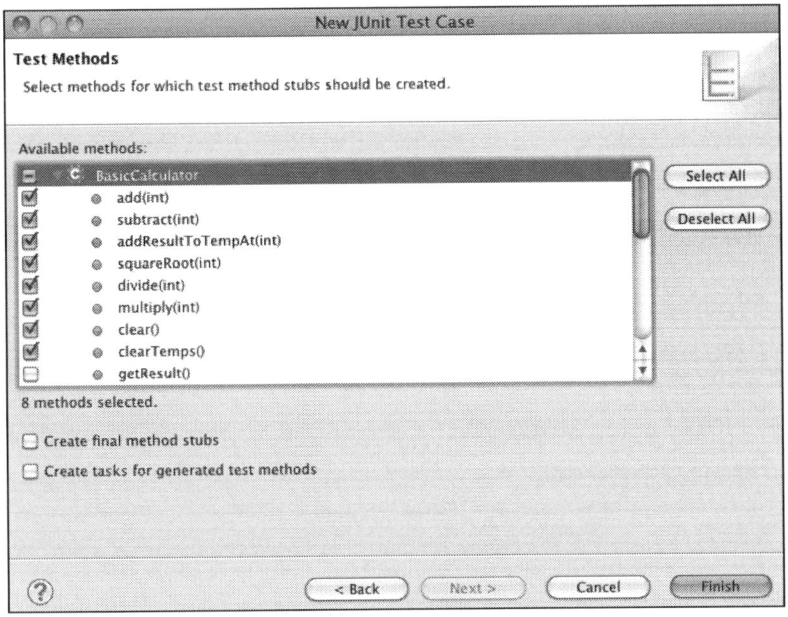

When you click on **Finish**, if you have not added JUnit to your project build path yet, Eclipse will prompt you for adding JUnit 4 to the build path (as shown in the following screenshot). You should respond by clicking on **OK**.

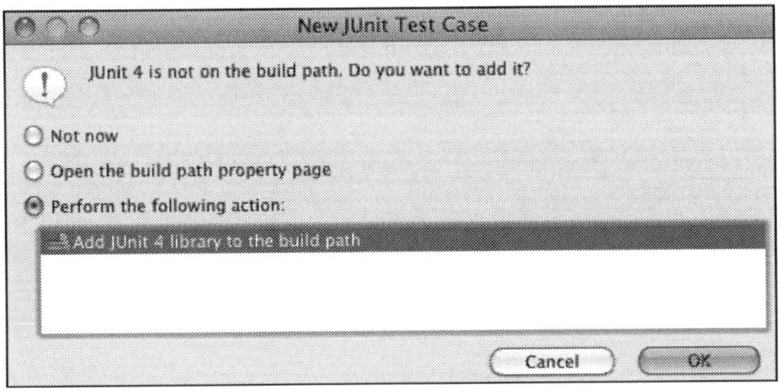

Eclipse will then generate the following code for you:

```
package chapter3.basiccalc.test;

import static org.junit.Assert.*;
import org.junit.Test;

public class BasicCalculatorTest {

  @Test
  public void testAdd() {
    fail("Not yet implemented");
  }

  @Test
  public void testSubtract() {
    fail("Not yet implemented");
  }

  @Test
  public void testAddResultToTempAt() {
    fail("Not yet implemented");
  }

  @Test
  public void testSquareRoot() {
    fail("Not yet implemented");
  }

  @Test
  public void testDivide() {
    fail("Not yet implemented");
  }

  @Test
  public void testMultiply() {
    fail("Not yet implemented");
  }
```

```
    @Test
    public void testClear() {
      fail("Not yet implemented");
    }

    @Test
    public void testClearTemps() {
      fail("Not yet implemented");
    }
  }
```

As you can see, Eclipse generated a lot of boilerplate code automatically. To be more specific, Eclipse generated all test methods (one for each method of our class under test), annotated then with `org.junit.Test` (`@Test`), and statically imported the assert methods.

The `@Test` annotation is used to indicate to JUnit that the annotated method should be run as a test case. Prior to running each test case, JUnit creates an instance of the class and then invokes the test case method on that instance. Later in this chapter, we will see how JUnit allows programmers to take advantage of this characteristic to initialize the context for test method that needs objects created before they can be executed.

If a test throws an exception, JUnit reports it as a failure. As you might have already figured out, all test method stubs generated by Eclipse are failing test methods. In fact, our test methods are failing because they invoke the `fail()` method from `org.junit.Assert`, which signals the failure of a test method by throwing an `AssertionError`.

In order to execute a JUnit-based test class, right-click on it and navigate to **Run As | JUnit Test**, as shown in the following screenshot:

Following the approach shown just now, try to run `BasicCalculatorTest`. The results of the tests are shown in the following screenshot in the **JUnit Runner** view. Whenever you run a JUnit-based class, this view is shown in the current perspective. This view shows you a list of failures and the test suite as a tree. Failing tests are indicated with a red bar. (as shown in the following screenshot, we have eight failing tests). The next sections provide more information on each toolbar of this window.

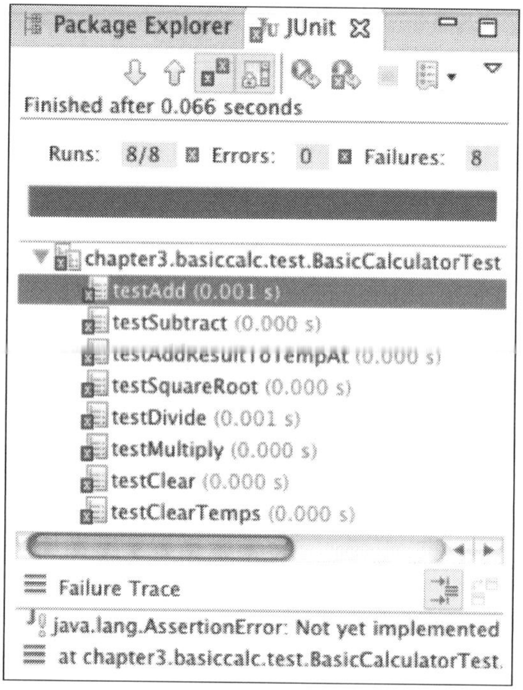

By now, you should be used to the fact that you can perform the same operation in Eclipse in different ways. There is no exception when it comes to running test cases. Here are some of the ways you can run your test cases:

- Let's say you want to run a single test method. Right-click on that test method from the Outline or Package Explorer, and navigate to **Run As | JUnit Test**
- To re-run a single test, choose a test in the **JUnit Runner** view, right-click on it, and select **Run**
- To run all the tests inside a project or a package, select the package, right-click on it, and navigate to **Run As | JUnit Test**

In order to check out what a passing test looks like, comment the body of the method `testAdd()` as in the following code snippet, and using the approach described just now, run only `testAdd()`:

```
@Test
public void testAdd() {
  //fail("Not yet implemented");
}
```

You should see a green bar; as illustrated in the following screenshot. To be more specific, it shows that `testAdd()` is a passing test method and the other test methods were not executed.

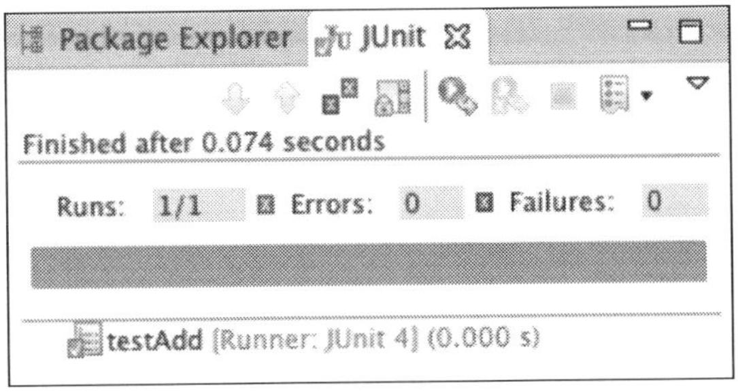

Before implementing tests that really evaluate the behavior of our calculator, it is worth taking a closer look at the **JUnit Runner** view, because you are going to be interacting with it throughout this chapter.

An overview of the Eclipse JUnit Runner view

After executing the JUnit-based classes, the JUnit Runner view is shown. In addition to indicating whether the recently run tests succeeded or failed, this view also shows why the failing tests failed and some useful toolbar commands. The JUnit Runner view has two panes, the tests pane and the failure trace pane. The tests pane lists all the executed tests as a tree. Upon selecting a failed test, the failure trace pane shows information on why the selected test failed. The following table presents a description of each toolbar command in the tests pane:

Icon	Description
	Moves the selection to the next failing test
	Moves the selection to the previous failing test

Icon	Description
	Shows only failing tests, if any
	Keeps the test list from scrolling
	Re-runs all tests
	Re-runs only all failing tests
	Stops the current execution
	Shows test run history

The following table describes the options available in the **Failure Trace** pane. To navigate from a certain failure to the related source code, double-click on the corresponding line in the **Failure Trace** pane.

Icon	Description
	Filters stack trace by removing unwanted stack frames from it
	Compares the actual and the expected results of String comparisons

Creating test cases

So far, we have not typed much; Eclipse has done everything for us. However, now it's time for us to do some typing. Aimed at implementing our test cases, we are going to take small steps. First, we need to comment out test cases that have not been implemented yet. It turns out that JUnit provides a better way to disable unfinished test methods, by annotating them with `@Ignore`. Methods annotated with `@Test` and `@Ignore` will not be executed by JUnit at all. JUnit treats them as if they were commented out. When annotating a method with `@Ignore`, you can also provide a string that explains why the underlying test method is being ignored.

In order to use `@Ignore`, you have to import `org.junit.Ignore`. Once imported, `@Ignore` can be used to annotate the methods, as illustrated in the following code. In the following code, we are telling JUnit to ignore `testSubtract`, because it has not been properly implemented yet. Go ahead and mark all methods but `testAdd()` with `@Ignore`. Executing our test suite now will result in no failures.

```
@Ignore("Not implemented yet")
@Test
public void testSubtract() {
   fail("Not yet implemented");
}
```

Now that the other test cases won't get in our way, let's start implementing `testAdd()`. Fortunately, writing unit tests is very straightforward. You just need to call chunks of your application's code, get the results back, and then check if they are exactly what you expected. Instead of having to write lots of if statements to compare the results when using JUnit, you can rely on the assert methods. Using assert methods, you specify the expected outcome, and then pass the actual result from invoking your application code.

For example, we can implement our `testAdd()` method by performing simple addition operations as shown in the following code. First, we need to create an instance of `BasicCalculator`. Second, we invoke the chunk of code that we are trying to test, namely, the `add()` method. Finally, we use an assert method to evaluate the result of invoking the `add()` method by passing `2` as parameter twice. The `assertEquals()` method expects `getResult()` to return `4`. Upon running our test case, we can see that the method under test seems to be working properly since `getResult()` returned `4`, as expected. Here is how we implemented `testAdd()`:

```
@Test
public void testAdd() {
  BasicCalculator calculator = new BasicCalculator();
  calculator.add(2);
  calculator.add(2);
  // result should be 4
  assertEquals(4 /*expected*/, calculator.getResult() /*actual*/);
  calculator.clear();
}
```

> Usually, writing unit tests should be easy. If writing them is taking too long, it might mean that your design needs some improvement; you should think about refactoring your code.

Before implementing more tests, a few more words about assert methods are in order. As you can see in the previous example, `assertEquals()` takes two parameters, the expected result and the value to check against the expected result. In our code, we are referring to this method through static import (check out the static import statement that Eclipse autogenerated for us). If we were to use `assertEquals()` without statically importing it, we would have to do more typing: `Assert.assertEquals(4, calculator.getResult())`.

The test for subtraction operations is similar to the one for addition operations. Again, we create an instance of the class under test, invoke the chunk of code under test, and check if the invoked code yielded the expected outcome. Go ahead and implement `testSubtract()` as shown in the following code snippet. Note that our `testSubtract()` method relies on `add()` to set the stage for `subtract()`. More specifically, `add()` is invoked to store the value `35` into the calculator's result variable. Then we subtract `5` from the result and use an `assert` method to evaluate whether after performing these arithmetic operations the result contains the value `30`.

```
@Test
public void testSubtract() {
  BasicCalculator calculator = new BasicCalculator();
  calculator.add(35);
  calculator.subtract(5);
  assertEquals(30, calculator.getResult());
  calculator.clear();
}
```

As you can see, all tests start by creating an instance of `BasicCalculator` (the one upon which the methods under test are invoked) and conclude with an invocation to `calculator.clear()` (in order to clear the value stored in result, so that it will not tamper with the results yielded by the other test methods). It is always a good idea to eliminate duplicated code from both your application and test code. We can extract this duplicated code into a method that does all the initialization and configuration of the test environment, and a method that contains the cleanup code. In JUnit 4, in order to have a method executed before each test case, simply denote it with `@Before` (`org.junit.Before`). Here is how we can remove part of the duplicated code:

```
private BasicCalculator calculator;

@Before
public void setup() {
  calculator = new BasicCalculator();
}

@Test
```

```
public void testAdd() {
  calculator.add(2);
  calculator.add(2);
  // result should be 4
  assertEquals(4/*expected*/, calculator.getResult()/*actual*/);
  calculator.clear();
}
```

Note that `calculator` is now an instance variable, and neither `testAdd()` nor `testSubtract()` (not shown in the code snippet) need to instantiate a local version of `BasicCalculator`; they simply use the `calculator` instance variable. Also note that all initialization code was moved to the `setup` method, which JUnit makes sure to invoke before executing each test method. Normally, methods that contain the initialization code needed for exercising the program under test are named `setup`. Likewise, methods containing the cleanup code are named `teardown`. Given that each test case now uses its own instance of `BasicCalculator`, if `result` were not a static variable, we would not need to invoke `clear()` on those instances. So, we would be able to safely get rid of all calls to `clear()`. However, since it is a static variable, we still need to invoke `clear()` after each test method.

> Since methods noted with `@Before` run before each test case, there are no side effects among test runs.

In our example, the cleanup code should be extracted into a method denoted with `@After`. By doing so, this method will be executed after each test case. Go ahead and implement the cleanup code as follows:

```
@After
public void teardown() {
   calculator.clear();
   calculator = null; //this isn't really necessary
}
```

Looking closely at our code, we can see that we do not need a new `BasicCalculator` object before executing our test methods. We can re-use one instance of `BasicCalculator` as long as we invoke `clear()` on that instance, after the execution of each test method. You might be wondering how to achieve this using JUnit. Actually, it is quite straightforward. JUnit provides an annotation that allows us to signal what methods must be executed prior to running all the test methods. The annotation that makes this possible is `@BeforeClass`. To be annotated with `@BeforeClass`, a method must be declared as public as well as static, and it can take no parameters. We no longer need our previous `setup()` method, so let's tweak it so that it will be executed only once for all the test cases (note that our `teardown()` method is still invoked after each test method to reset the result of calculator):

```
private static BasicCalculator calculator;

@BeforeClass
public static void setup() {
  calculator = new BasicCalculator();
}

@After
public void teardown() {
  calculator.clear();
}
```

Now, let's move on to more challenging arithmetic operations. Let's get down to tackling the test of the `multiply()` and `divide()` methods. Luckily for us, testing multiply is straightforward, as shown in the following code snippet:

```
@Test
public void testMultiply() {
  calculator.add(10);
  calculator.multiply(3);
  assertEquals(30, calculator.getResult());
}
```

Although `testMultiply()` looks quite simple, upon running it, we discover a bug. As shown in the following screenshot, instead of returning `30`, `getResult()` returns `9`. Note how the error message uses the expected value and the actual value to inform us that our test failed.

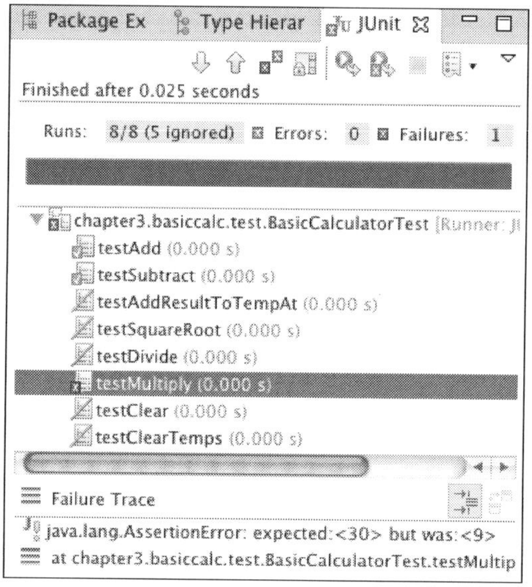

Puzzled, we take a look at the implementation of multiply to discover that rather than multiplying `result` by the parameter `n`, we are multiplying `n` by itself and assigning this value to `result`. Fortunately, the fix is easy:

```
public void multiply(int n) {
  result *= n;
}
```

As you can see, even simple test methods are effective weapons against the bugs that lurk in our code. Nevertheless, so far we just wrote test methods that exercise the happy path of the code of our `BasicCalculator`. Sometimes we need to write test methods that exercise edge cases of the chunk of code under test. In Java, inputs that cause a certain method to throw an exception can be considered edge cases. For example, an edge case that our calculator has to cope with is division by zero.

In Java, an attempt to divide an integer by zero will lead to an `ArithmeticException`. If it is expected from our calculator to throw an `ArithmeticException` whenever there is a division by zero, we need to explicitly state this expectation in our tests.

Fortunately, JUnit provides a way to deal with these edge cases and the associated expected exceptions effectively. The `@Test` annotation supports two optional parameters. The first, `expected`, specifies that a test method should throw an exception. If the underlying method throws no exception or the thrown exception is different than the one specified, the test fails.

For instance, we can use this annotation to implement an edge case of our `testDivision()` method as in the following code. It specifies that the method `divide()` from `BasicCalculator` should throw an exception (`ArithmeticException`) whenever a division by zero takes place. Note that we do not need to use any assert method since the only outcome that we are expecting is an `ArithmeticException`. It is worth mentioning that the class passed as parameter to `expected` must be a subclass of `java.lang.Throwable`.

```
@Test(expected=ArithmeticException.class)
public void testDivide() {
  calculator.add(10);
  calculator.divide(0);
}
```

There is yet another way of achieving the same behavior. You can go about declaring expected exceptions through JUnit's `ExpectedException` rule. First, let us give you a crash course in JUnit rules.

In earlier versions of the test framework, it was not possible to modify or augment JUnit's internal knowledge about a given test suite. Using the current version, we can use rules to modify test results, change a test in a certain way before running it, and add information to test results. We will use the `@Rule` annotation to inform JUnit which rules to apply to the underlying test.

By using `ExpectedException` in conjunction with `@Rule`, you can specify in-test expectations regarding the exceptions that should be thrown for a particular test method. It takes more typing than expected, however, as shown here, you have more control over what is expected. For instance, you can even specify the message what the expected exception should contain.

```
@Rule
public ExpectedException thrown = ExpectedException.none();

@Test
public void testDivide() {
  thrown.expect(ArithmeticException.class);
  thrown.expectMessage("/ by zero"); //specify the expected message
  calculator.add(10);
  calculator.divide(0);
}
```

As for the second optional parameter of `@Test`, `timeout`, it causes a test method to fail if it takes longer than a specified amount of milliseconds to execute. To give you an example, let's say that we implemented `testSquareRoot` as an infinite loop, which fails after 1 second. Upon running the code below you should get a red bar along with a `java.lang.Exception` containing the following message: "test timed out after 1000 milliseconds".

```
@Test(timeout=1000)
public void testSquareRoot() {
   while(true);
}
```

Okay, by now, we believe that you got the gist of writing test methods using JUnit and Eclipse. We even were able to uncover a bug in our calculator. Granted, it was a very simple, easy-to-spot bug.

We are sorry to say that not every bug is simple to uncover. Sometimes you might get a red bar that leaves you at a loss to figure out how to fix the problem. In other words, when a test fails, and you just cannot figure out why, even after spending quite some time staring at the code, you need to turn to a debugger. To illustrate this somehow, let's implement the test method, `testAddResulToTempAt`. The `addResultToTemp` method should perform the following operation: add the current value of store in the variable result to a position in the `tempVars` array, according to the index passed as parameter. During the implementation of this test method, we realized that we did not implement a method to get this array of temporary results back. When implementing unit tests, realizing that you forgot something happens more often than not. To set this matter right, we add a simple get method, named `getTemps`, which simply returns the variable `tempVars`.

```
public int[] getTemps() {
  return tempVars;
}
```

As for our test method, which uses our recently implemented `getTemps`, we implemented it as the following:

```
@Test
public void testAddResultToTempAt() {
   calculator.add(9);
   calculator.addResultToTempAt(1);
   int[] temps = calculator.getTemps();
   assertEquals(9, temps[1]);
}
```

As you can see, we once again use add to store the value `9` into the variable `result`. Afterwards, in order to store `9` into the slot number `1` of the temporary memory, we invoke `addResultToTempAt` and pass `1` as parameter. The expected behavior is that the array that represents the calculator's temporary memory now contains the value `9` at index `1`. However, the test fails (as shown in the following screenshot). Instead of the expected behavior, we get a `NullPointerException`. The only clue we got from observing the **Failure Trace** pane is that the exception is thrown at line number 16 of our `BasicCalculator` class. Just by looking at this line there is no way you can guess what is wrong (well, it also depends on your Java experience). It is for problems like this that the debuggers exist. So, we will not spoil the fun of discovering the root cause of this problem. Rather, we are going to go over the tools of the trade. That is, we are going to explain how to use Eclipse built-in Java debugger so that you can find and fix this bug by yourself.

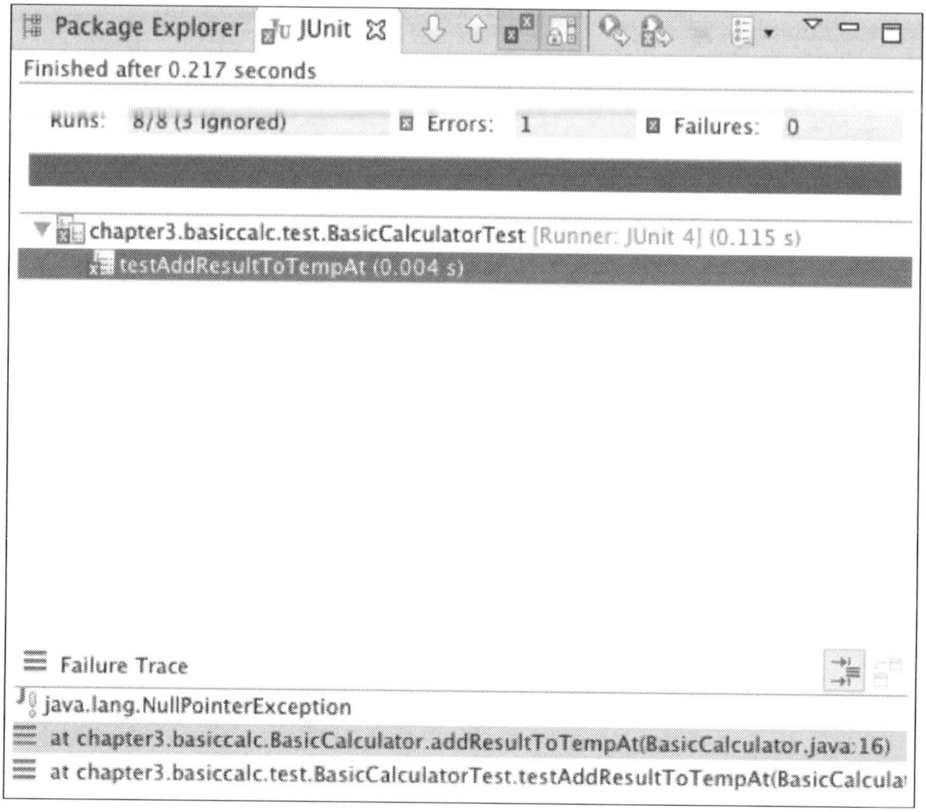

Debugging when problems arise

If you have been a programmer for a while, chances are that you are familiar with the classic debugging technique known as trace code. In case you are not familiar with trace code, let's break it down for you. In this debugging technique we introduce `println` or `print` statements throughout the code under inspection in order to show information regarding variables or whether the program has taken a certain path. With this definition in mind, you might be wondering, why would you need a debugger? The answer is simple and twofold. First, using a debugger you can achieve the same results without too much fuss. When applying trace code, you need to include print statements, recompile the program (or the particular chunk under inspection), run, and analyze the output of the trace code. Also, after finding the bug, you have to remove all trace code. Without mentioning that whenever a bug appears, you have to repeat the procedure all over again. Trace code is a cumbersome technique. In contrast, when using a debugger, achieving all that is just several clicks away. Second, you can get a lot more from debuggers; they allow you to peruse aspects of your program that would be quite difficult to figure out just by looking at the output.

As we mentioned, Eclipse ships with a full-blown debugger. This built-in Java debugger provides all debugging support functionality; it is possible to execute programs step-by-step, set breakpoints, inspect variables and values, and pause and resume threads.

There are many ways to launch a Java program in the debug mode. One option is to select the Java class file with the main method that you want to debug in the **Package Explorer** view and then click on the **Debug** [] button in the workbench toolbar (or select **Run** | **Debug** from the workbench menu bar). It is also possible to achieve the same thing by selecting **Run** | **Debug as** | **Java Application** from the workbench menu bar, or by clicking on **Debug as** | **Java Application** in the **Package Explorer** pop-up menu or in the drop-down menu on the **Debug** toolbar button. During debugging activities, Eclipse switches to the Debug Perspective, but before switching to it, the window shown in the following screenshot asks whether you want to change from the current perspective to the Debug Perspective.

> Run configurations and debug configurations are shared, that is to say, any run configuration that you used to run your application should be available in the debug-related menu.

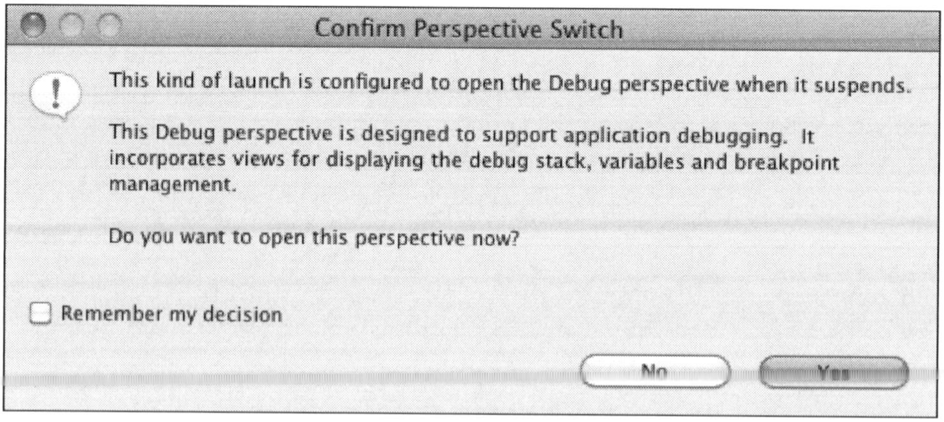

When you click on **yes**, the Debug Perspective appears, as illustrated in the following screenshot. This perspective allows you to control all threads of execution of the program under inspection. In the Debug Perspective: the call stack in the top-left corner indicates the methods of your program that were invoked, in the top-right corner you can manage the breakpoints and inspect the content of variables. These views are updated as you interact with your running program by stepping through its execution and adding and removing breakpoints.

The basic way of interacting with the debugger is by setting breakpoints. A breakpoint is a way of indicating to the debugger the line of code where you want the execution to pause. To set a breakpoint, while in the Java perspective, select the source code file you want to debug and open it in the editor (by double-clicking). Skim through the code and place your cursor on the marker bar, which is along the left-hand side edge of the editor area, on the line that you want execution to stop. Then, double-click to set the breakpoint. Alternatively, you can right-click on this position and select **Toggle Breakpoint**.

> Apart from regular breakpoints, there are conditional breakpoints. Conditional breakpoints are activated when the Boolean expressions associated to them evaluate to true.

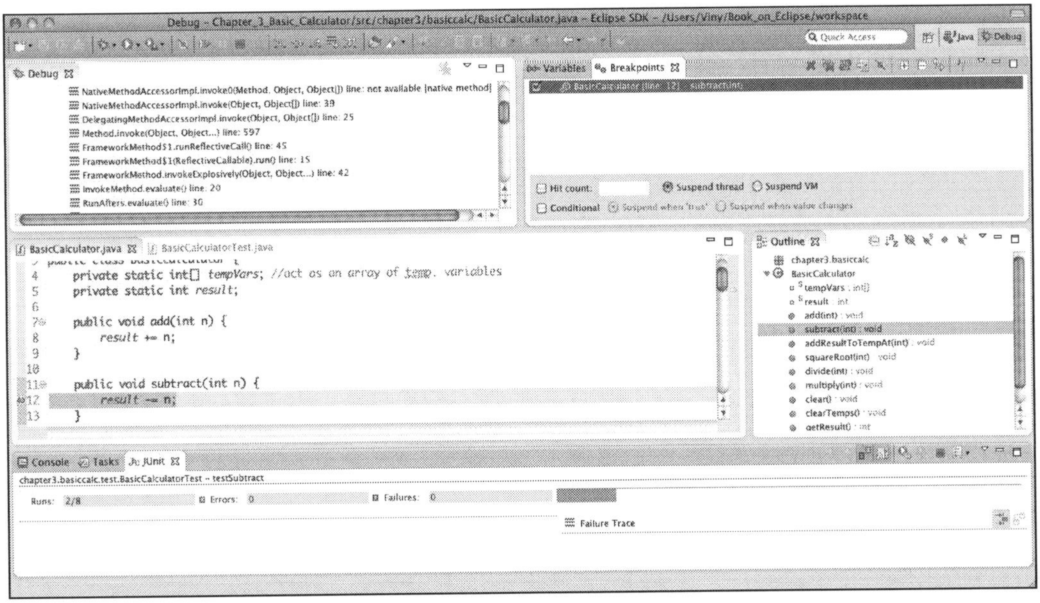

Once a breakpoint is hit, Eclipse pauses the execution, and the line containing the breakpoint is highlighted, which gives you a visual feedback of the next line that will be executed. This gives you a chance to examine the state of variables, and so on.

> If you add code before or after a line containing a breakpoint, Eclipse makes sure to move that breakpoint along with its respective line. So you can add code without having to worry about changing the positioning of breakpoints.

To resume execution after hitting a breakpoint, click on the **Resume** button in the **Debug** view's toolbar. This button makes the program execute until completion or proceed until it reaches the next breakpoint.

> You can add breakpoints even during a debug session; you do not need to suspend or stop the execution of the program. Also, bear in mind that since you cannot set more than one breakpoint per line, it is important not to put more than one statement on a single line. If you do so, you will not be able to examine these lines properly. It is also worth mentioning that if there is no breakpoint, execution will not pause and your program will execute normally, as if you had run it in the Java Perspective.

Aimed at inspecting the behavior of your program, Eclipse Debugger allows for stepping into the code line by line through the program execution. This can be done through two commands: step into (*F5*) and step over (*F6*). The difference between these two commands is only noticeable when the next instruction is a method call. Selecting step over makes the debugger execute the method (without entering it) and proceed to the next statement following the method call. It is worth mentioning that even though the debugger does not show the method's body, it will be executed normally. On the other hand, if you choose the step into command, the debugger will proceed to the first line of the underlying method.

After stepping into a method call it is still possible to skip the rest of it: use the step return command (*F7*). The debugger will proceed to the next line after the method call. The following table gives an overview of the main commands in the **Debug** toolbar.

Icon	Description
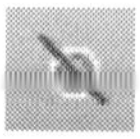	Disable all breakpoints.
	Resume execution.
	Stop the current debug session by terminating the current execution.
	Proceed to the next statement of the program under inspection. If the next statement is a method call, this command will step into the method body code.
	Proceed to the next statement of the program under inspection. If the next statement is a method call, this command will execute the method body without entering it.
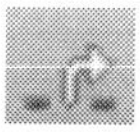	Exit the current method and go to the calling code.
	Re-enters the selected stack frame in the Debug view. Since it relies on the existence of an stack frame, it is not possible to use this option when debugging native methods.

As you step through your program execution, it is also possible to peruse and modify the state of your program. You can examine the contents of local variables, parameters, and visible fields by looking at the **Variables** window. It is also possible to investigate the contents of a variable by hovering the mouse pointer over it in the Java editor. As for non-primitive variables, they can be expanded in order to examine their individual elements. Apart from inspecting the value of variables, you can also examine the value of Java expressions in the editor in the Debug Perspective. Select an expression, then right-click on it, and choose **Display**. Eclipse evaluates the selected expression, taking the current scope into account, and shows the result in a pop-up window.

During a debug session, you can also modify the value of variables. To do so, select the variable that you want to modify in the **Variables** view; this will make its value appear in the bottom-half window of the **Variables** view, so that you can edit it. After editing, right-click and choose **Assign Value**.

Let's say that during your debug session you found a bug, and luckily, you also know how to fix it right away. Do you need to stop the debug session to start fixing the program? The answer is no. Thanks to a feature called hot code replace, editing and debugging can be carried out in tandem; you can edit your program without having to stop your debug session. The only side effect is that after modifying a method, Eclipse has to throw away the previous stack frame and begin the method again from the first line.

> It is likely that you have been using a Java virtual machine 1.5 or later, but we should warn you that hot code replace is only supported by Java virtual machines later than 1.4.2.

We covered the main functionalities of the debugger, now, armed with this debugging knowledge, go back and try to pinpoint the problem with `testAddResultToTempAt`.

Summary

In this chapter, we presented the basic of creating unit tests using Eclipse and JUnit. You saw how Eclipse is able to generate test method stubs and automate the execution of these methods by seamlessly integrating the JUnit. Moreover, we also gave an overview of the main JUnit features as, for example, the most commonly used annotations. Apart from unit tests, this chapter also briefly covered how to use the Java built-in Eclipse debugger.

Keep in mind that there are some testing-related topics which we did not cover in this chapter. Earlier on in the chapter, we mentioned that tests need to be independent from each other so that you can re-run them in any order that you wish. Sometimes, however, objects interact with other objects, which makes the creation of independent test methods more complex. In a unit test, when there are objects that may compromise the independence of the test or whose complexity may make the use of the real objects impractical, mock objects are used. These mock objects simulate the behavior of real objects. Luckily, there are frameworks that automate most of the process involved in creating mock objects. We encourage the interested reader to read about Mockito (`http://code.google.com/p/mockito/`).

Apart from a good set of test methods, whenever you set out to develop a reasonable-sized application, you need to have a back-up strategy. Usually, for code projects, this strategy includes using a version control program. A program that keeps track of different versions of programs is called version-control system. The next chapter describes an example of version-control system, namely, Git.

4
Version Control Systems

Source code management, or **version control**, is the practice of managing the revisions of a code with comments so that the code can be retrieved easily. The concept of **version control systems** (**VCS**) has been around for a long time. The first implementations were file-oriented, centralized, and without network access capability.

An evolution has come with **Concurrent Versioning System** (**CVS**) that was explicitly designed for collaborative development and had a merging mechanism instead of a locking-based one. The latest generation presents a major concept modification, that is, a complete decentralization.

This clearly seems more appropriate, for nowadays network-based systems are likely going to be the most used ones in the near future. A good versioning system should at least offer the following features:

- The possibility to add, remove, rename, or delete files
- The possibility to find which files have been altered by which user and when
- The provision to keep track of several versions of a product
- The possibility to go back and forth among versions
- The possibility to merge code
- The provision to revert a modification (go back in history)

In this chapter, we are going to present the main characteristics of VCS and how to make use of the three most popular version control systems inside the **Eclipse** platform, that is, CVS, **Subversion** (**SVN**), and **Git**.

CVS

The development of a product is usually performed by several software engineers. Version control systems such as CVS are typically used to manage the files related to the development. Although version control systems are generally applied on programming projects, they are not limited to that purpose and can be used for any task that involves the management of datafiles.

CVS is based on the concept of branches, where each developer can commit his modifications to the same set of resources. The main branch, **HEAD**, is generally used when there is a certainty that the implementation is stable. It is possible to use several branches and merge the files to the main branch at any moment.

CVS repositories maintain a history of every committed change made to the resources over time. A resource may be compared with or replaced with any prior existing revision.

Eclipse provides facilities for merging, committing, and accessing a project's history. In the following sections, we'll present some of the basic functionalities regarding the use of the CVS technology.

Repository

The first thing that needs to be done is to connect the Eclipse workspace to a **CVS repository**. This operation can be performed by navigating to **File** | **New** | **CVS** | **CVS Repository Location**. A window like the following screenshot will appear:

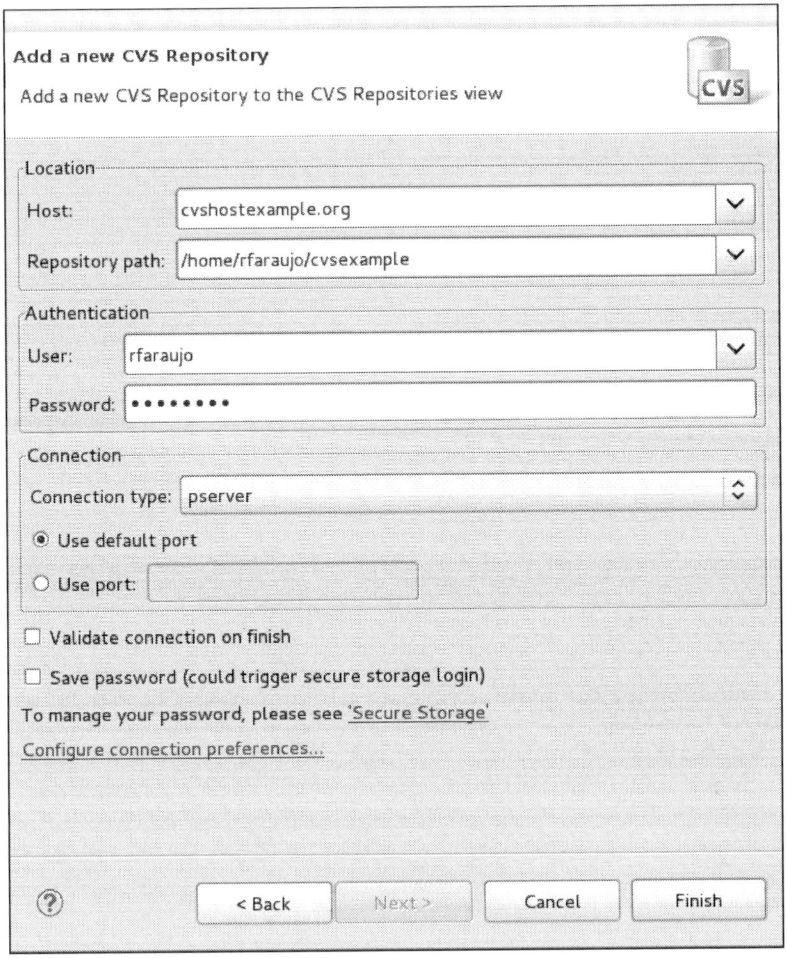

It is necessary to select the **Host**, **Repository path**, **Username**, and **Password** of the desired repository. The repository will be added to the workspace, and the following screenshot of the **CVS Repositories** view will be seen:

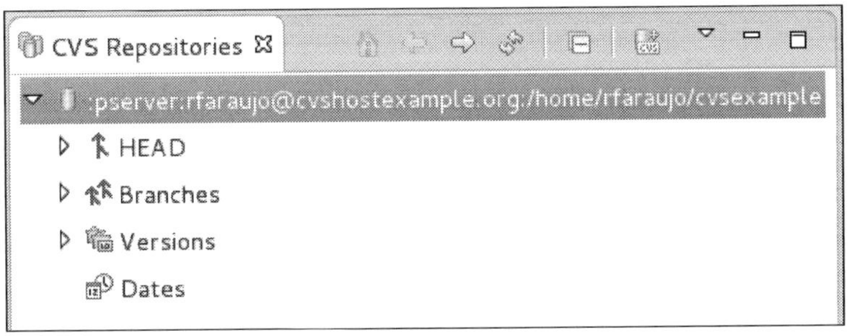

Checking out a project

In order to check out a project, expand the **HEAD** label, select the underlying project, right-click on it, and go to **Checkout**. This will load the project into the workspace.

It is also possible to use the wizard by navigating to **File** | **New** | **CVS** | **Projects from CVS**. You can select the **repository location** and from the second page, you can choose the list of projects with which you wish to work.

Synchronizing

While you are editing the resources of a project, there will be times when you will want to commit it back to the repository. For this purpose, you can click on the resource and select **Synchronize with Repository**.

A **synchronize** view will be opened. There will be three options in it:

- The **Incoming** mode, where the view shows only incoming changes
- The **Outgoing** mode, where the outgoing changes are shown
- **Both modes together**

To commit a change to the repository, you can navigate to the **outgoing changes** option and select **Commit**. To load any possible changes to your workspace, right-click on the **incoming changes** option and select the **Update** option.

It is possible that conflicts may occur when committing or updating; for example, let's assume you and another developer have modified the same resource. In that case, a **merge** tool can be used to select the version you want to keep.

Branches

When there is need to keep multiple versions of a product, for instance, it is recommended that a branch be created; this can be done by performing the following steps:

1. After right-clicking on the project, go to the **Branch** option.
2. In the window that follows, leave the **start working in the branch** box checked.

   CVS will automatically propose a version name that can later be used for merging purposes.

3. In the **CVS repository** view, it is possible to see the created branch under the **Branches** node.

SVN

SVN or Subversion, allows multiple developers to work on the same project. The project is usually maintained at the server, and the ones that are working on it do a checkout (creating a local copy) of the files from the server. It is also possible to commit the modified files back to the server. In order to perform any operation, there is need for connecting to the server.

SVN, in contrast to CVS, can track not only file change history, but all the changes regarding the shared projects, such as allowing the addition, deletion or removal of all resources without creating new items, and losing information about deleted ones. A commit operation in SVN is absolutely atomic, that is, the repository will never reach an inconsistent state, and versioned properties' support is available.

Eclipse does not come with an SVN client integrated by default. It is necessary to install an additional plugin, namely the **Subversive project**. Nonetheless, it is maintained as an Eclipse project in the Eclipse portal. Another option is to make use of the **Subclipse project** (`http://subclipse.tigris.org/`). The Subversive plugin aims to integrate the Subversion version control system with the Eclipse platform. The plugin allows the possibility of working with projects stored in Subversion repositories directly from the Eclipse workbench, in a similar fashion as other Eclipse version control providers, such as CVS and Git.

The project was started in 2005 by the Polarion Company as an open source project and was later migrated under the Eclipse umbrella as an official Eclipse project in 2007. Nowadays, the Subversive project consists of the Subversive plugin for Eclipse, developed as an Eclipse open source project, along with the Subversion connectors, developed as an external open source project.

Subversive has been designed to be used as a full-featured SVN client, so it is possible to update, commit, merge changes, work with SVN properties, view change history, and perform other operations with SVN directly from the Eclipse environment. Moreover, Subversive evolves together with the Subversion project to provide Eclipse users with the features that appeared in the new versions of the SVN implementation.

The **Juno version** of the plugin can be installed using the update-site `http://download.eclipse.org/technology/subversive/1.0/update-site-1.0.1/`.

In order to install it, simply navigate to **Help** | **Install new Software** and add the update site as shown in the following screenshot:

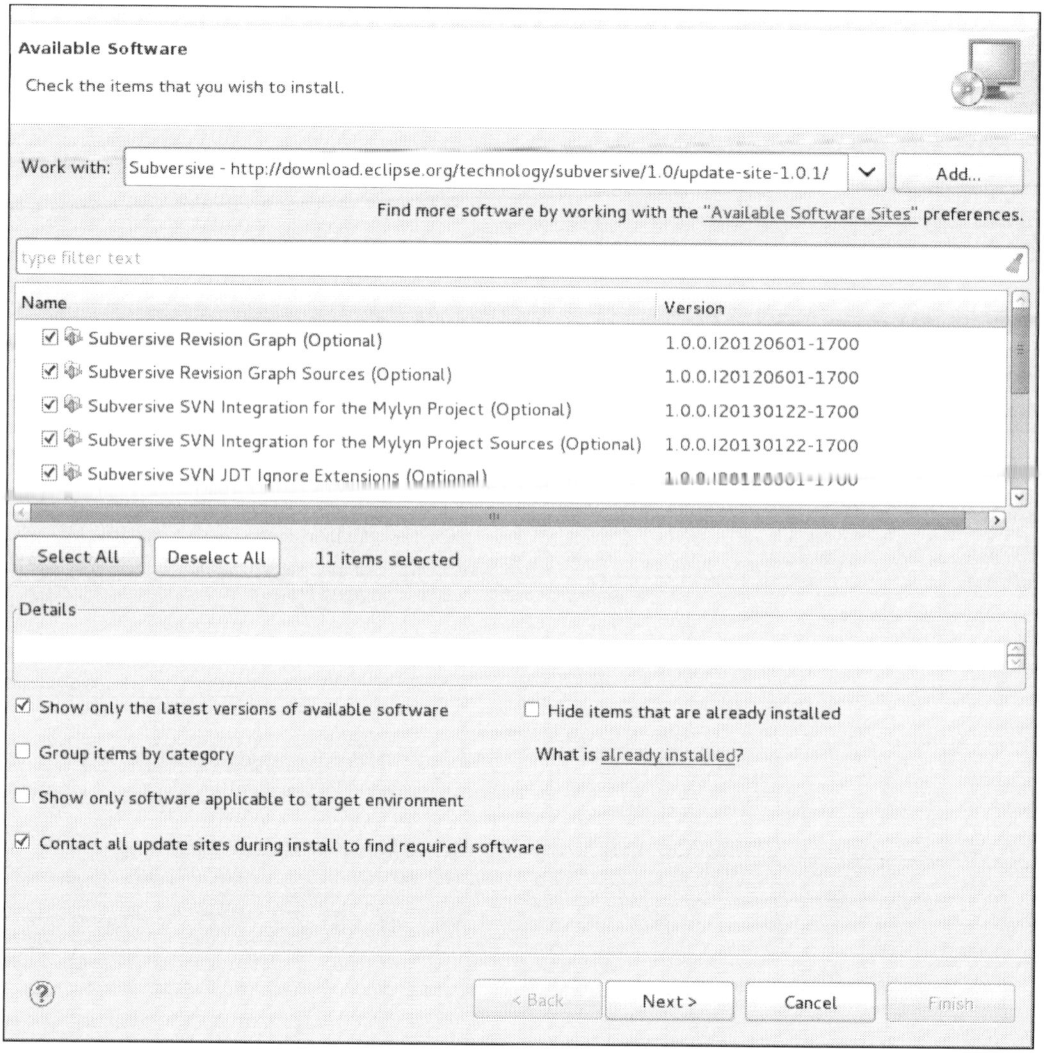

In order to exemplify the use of SVN, we are going to use the **Apache Hadoop repository** as the basis. In order to associate a repository with the Eclipse workspace, you need to navigate to **File** | **New** | **SVN** | **Repository Location** and fill the following screen with the server location `http://svn.apache.org/repos/asf/hadoop/common/`. Have a look at the following screenshot:

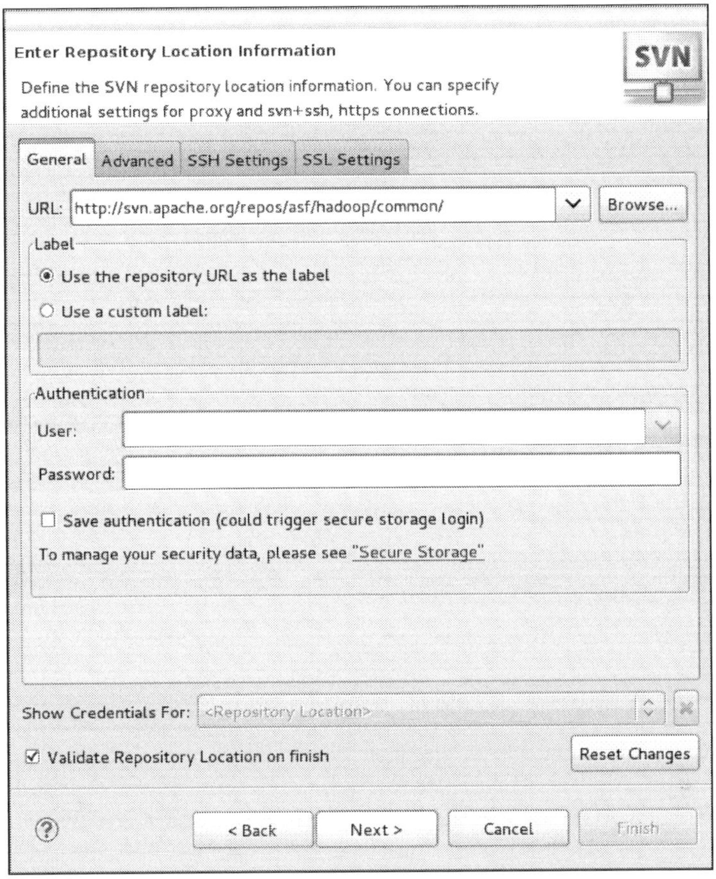

After you have clicked on **Finish**, the repository will be added to the **SVN Repositories** view, and it will be possible to view the **trunk**, **branches**, and **tags** nodes here. In the **trunk** node, you will see all the maintained files and early revisions. It is possible to select a start point by selecting a specific revision.

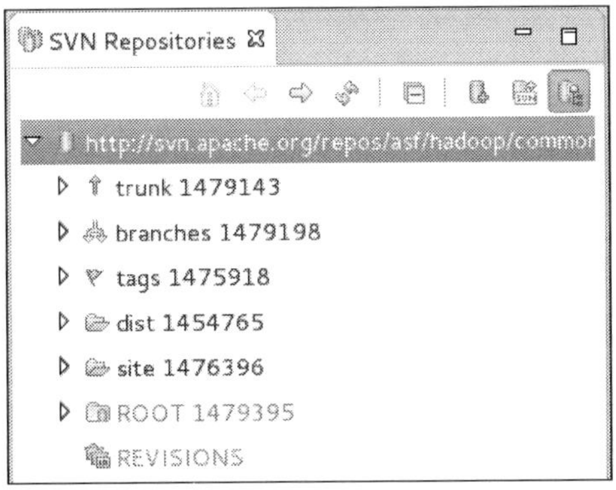

There may be more than one branch on a given version control system that can be selected in the **branch** node.

The **tags** are usually defined to make it easier to go to a specific revision of the product, relating it to a commit; for example, to specify a product version number.

The operations related to **Checkout**, **Update**, and **Commit** are pretty similar to what we have presented in the CVS section.

Git

Git works a little bit different from SVN. It does not copy the repository, but it clones it. In this context, it is possible to perform any operation locally, without a connection to the Internet, because the complete repository is available. It is possible to perform several actions, such as branching, merging, or committing.

A few differences between Git and SVN are as follows:

- Git is distributed
- Git stores the metadata instead of the actual files along with the existing commits
- Git can work without an active Internet connection since the operations are performed locally

Eclipse comes with a preinstalled plugin to deal with the Git program. The **EGit** is an Eclipse plugin that provides the Git version control system. As previously mentioned, because Git is a distributed source code manager, every developer has a full copy of all the history of every revision of the code, making the queries against the history very fast and versatile.

If there is need to install it, the user can navigate to **Help | Install New Software** and add the repository `http://download.eclipse.org/egit/updates` as it is shown in the following screenshot:

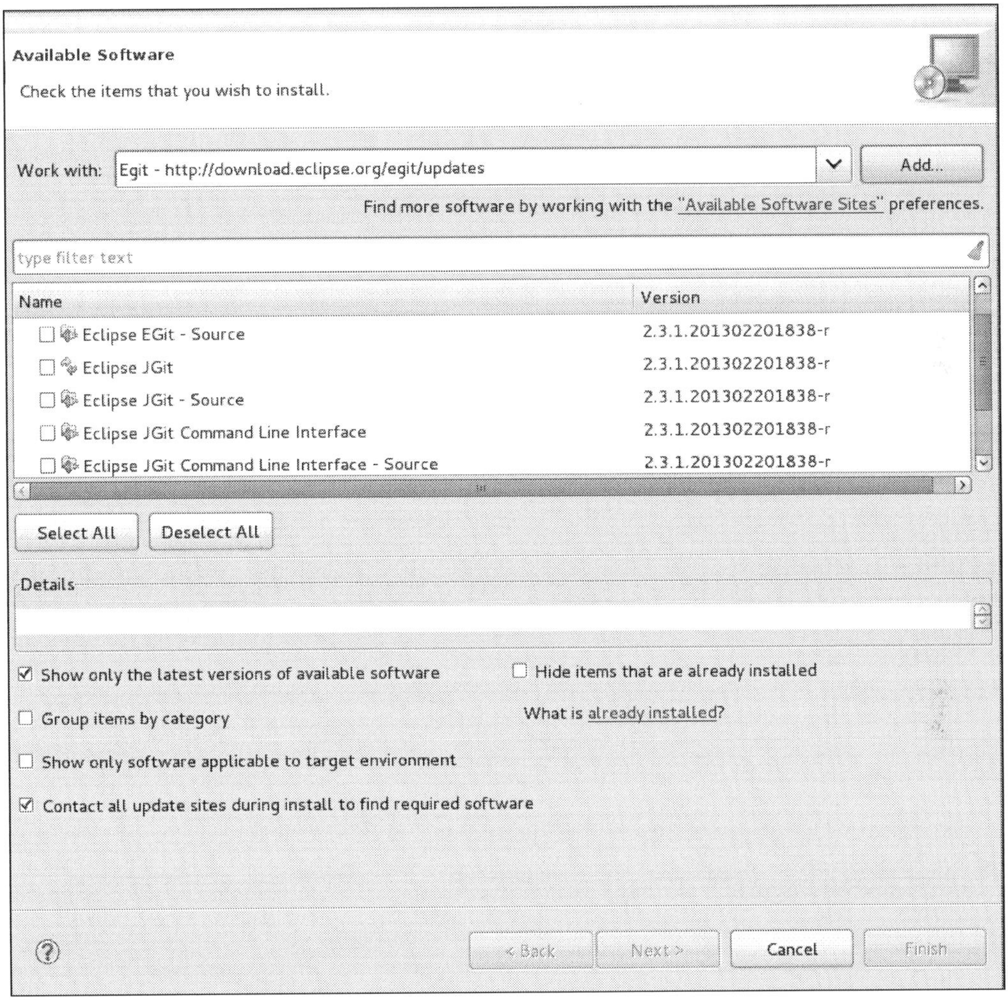

When dealing with the EGit plugin, it is possible to change the perspective to the EGit one. In order to do this, simply navigate to **Window | Open Perspective** and choose **Git Repository Exploring**.

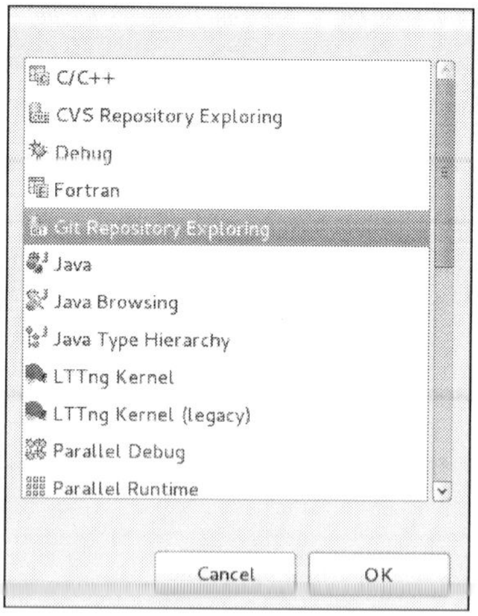

In order to create a **Git repository**, one can navigate to **File | New | Git | Git Repository**. It is possible to create a local repository if needed. The other option is to import files from an already existing project by navigating to **File | Import | Git | Projects from Git** as shown in the following screenshot:

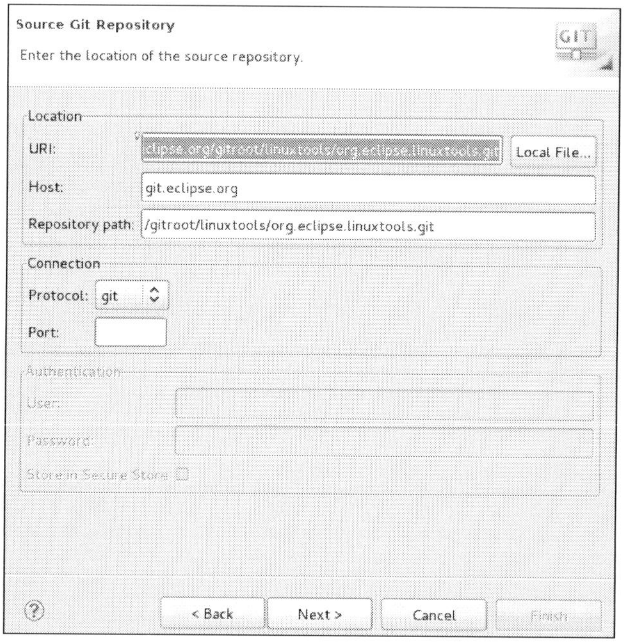

In this example, we use the repository of an Eclipse project and the Linux tools that can be cloned using the URL `git://git.eclipse.org/gitroot/linuxtools/org.eclipse.linuxtools.git`.

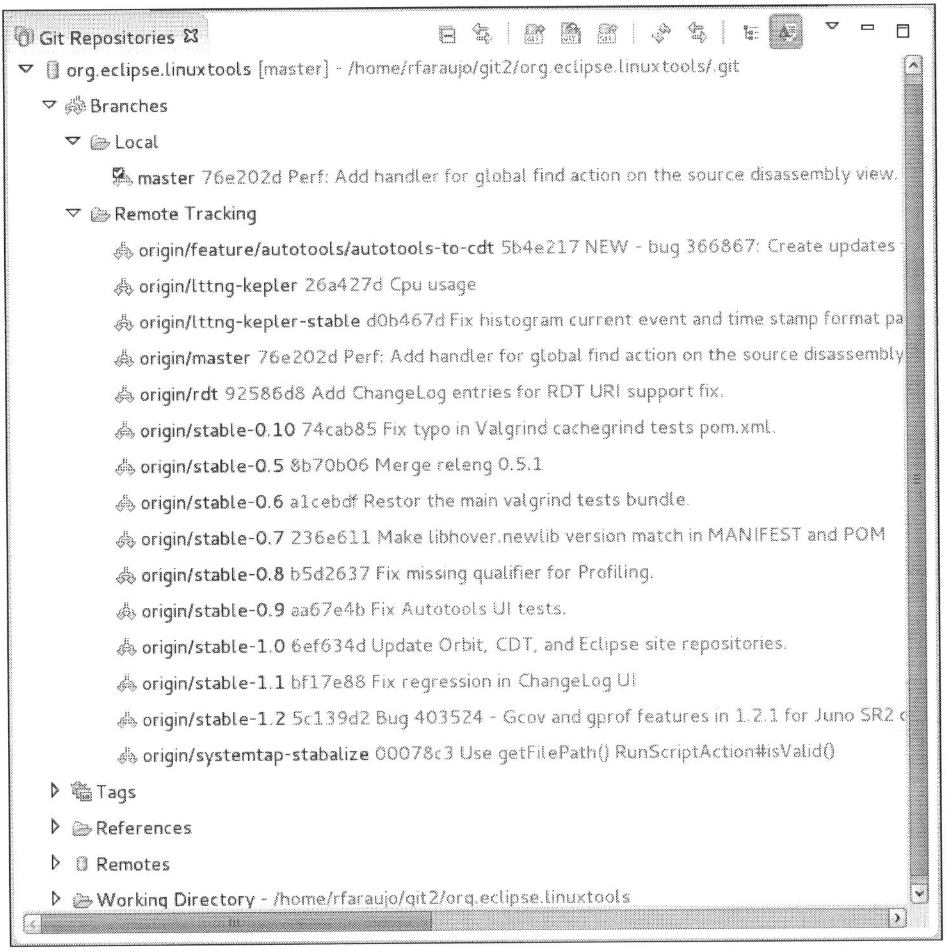

After the repository is cloned, the **Local** and **Remote Tracking** are shown in the **Git** perspective. The main **Branches** node is the master one; nonetheless, others can be created and visualized in the perspective in the **Branches** node. By supporting branching, Git provides support for you to work on different versions of your collection of resources in parallel. It is possible to work on a fix of a bug in a different branch and merge it back to the main one once this work is finished. **Branches** in Git are local. A branch created in a local repository, which was cloned from another repository, does not need to have a counterpart in the **remote repository**. **Local** branches can be compared to **Remote Tracking** branches that proxy the state of branches in another remote repository.

Like SVN and CVS, it is also possible to look at the history of the changes by going to the **History** tab shown in the following screenshot:

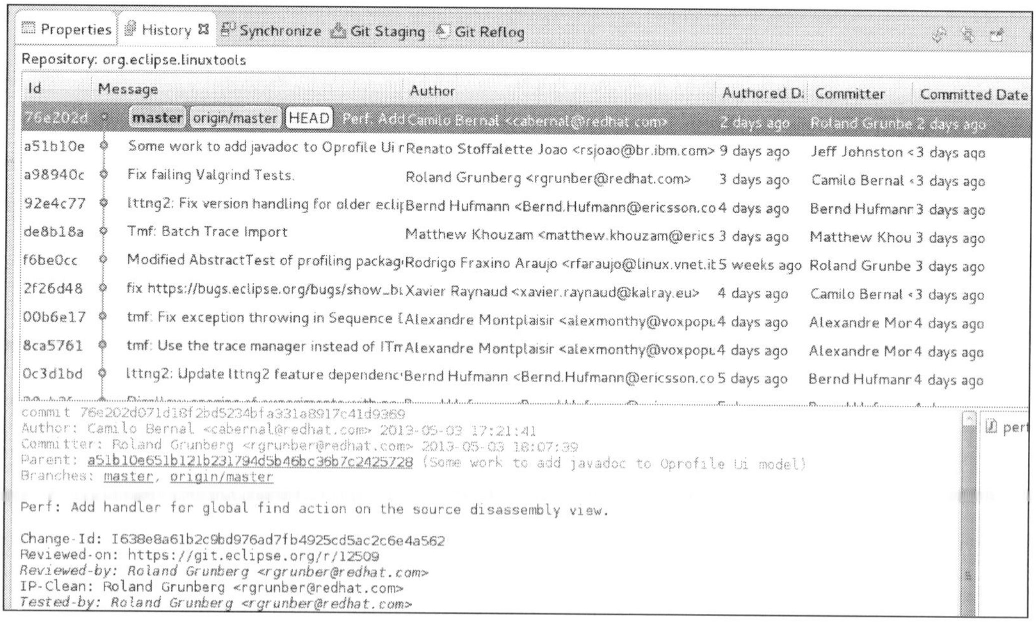

In the previous screenshot, it is possible to see the last commits, along with details of the **Author**, **Committer**, and **Committed Date**. If one selects **commit**, it becomes possible to visualize further information. It is also possible to move to the position of a specific commit, rebasing the resources in that position by right-clicking on it and selecting the **Rebase** option.

After checking out a given branch by right-clicking on it and going to the **Checkout** option, the project will be available on the Eclipse workspace. When you modify a given file, you often need to add that modification to a local commit so that it gets pushed to the remote repository. Usually before committing to a **Git repository**, you need to mark the changes that should be committed. This operation is called by adding the files to the staging area. Nevertheless, in the EGit plugin, you can select which file is going to be committed in the **GUI interface** by selecting the **commit** option when right-clicking on the branch being used. Afterwards, it is necessary to type a commit message describing what modification was performed.

After this operation, the **push** option can be used so that this modification gets pushed upstream. In some cases, when another person has also modified the same resource, there may be merge conflicts; they can be resolved with the assistance of the merge tool of EGit. The **commit** option can be seen when right-clicking on the node containing the current branch (**master**, in this case).

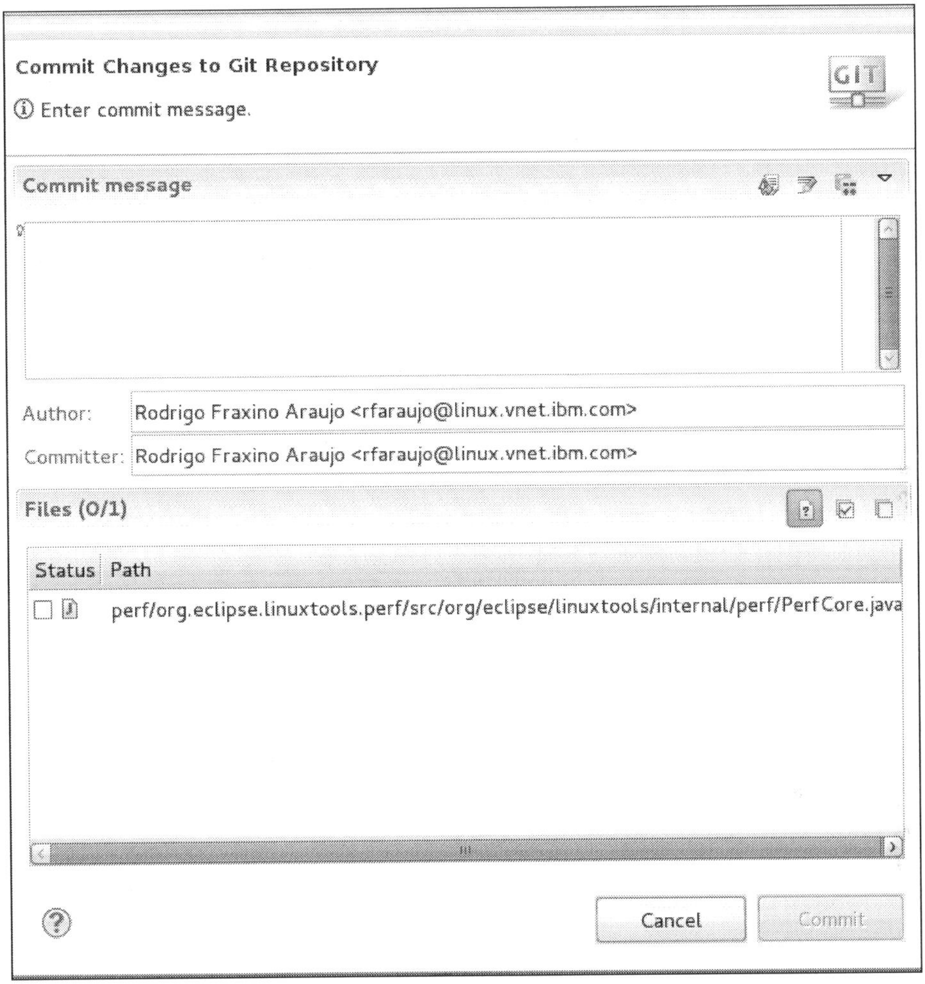

Another operation that can be performed is the checkout. This operation allows you to jump among the branches of a repository. In EGit, you simply need to right-click on the desired branch and select the **Checkout** option. Remember to verify if you have committed any changes that you performed in the previous branch, otherwise they may get lost if you stash (discard) them.

When updating the local repository with the **remote repository**, you need to use the **fetch** option. Afterwards, you can choose to **rebase** or **merge** the upstream repository with yours using the **rebase** option; this means that the repository is going to get cloned and the possible commits of yours that are not upstream will later be applied to the tree. The other option is to merge the files present in the local commits with the **remote repository**. If you choose to use the **pull** option, by default it performs the **fetch** and **merge** operations.

Summary

In this chapter, we have discussed a way to organize and keep track of modifications of source files that are usually necessary for the development of software projects. We have introduced the CVS, SVN, and Git version control systems that can be considered the most popular ones nowadays. We have given special attention to Git, which is a trend technology, because it is a recently introduced distributed option.

In the next chapter, we will discuss how to deal with graphical programming using one of the most-used Eclipse APIs: the SWT.

5
SWT

The **Standard Widget Toolkit** (**SWT**) consists of a set of tools created by the Eclipse project developers to build **Graphical User Interface** (**GUI**) applications easily. This open source widget toolkit provides efficient and portable access to the user interface facilities by focusing on the target operating system.

Although Java provides support for the development of GUI applications by using the **Abstract Windowing Toolkit** (**AWT**) and the Java foundation classes (swing), both cannot seamlessly integrate with the target operating system.

SWT supports widgets that makes use of operating-systems-specific libraries and grant access to them from a set of Java classes that are responsible for communicating with the native code. The outcome is the achievement of better performance with the native code, preserving the ability of writing the desired code only a single time so that it can be used across multiple platforms. Since the code is native, the look and feel of the applications follow the target platform, and the task of drawing the graphics is performed by the compiled code that is not part of the **Java Virtual Machine** (**JVM**). A minor setback is the necessity of deploying a run-time library for each platform that is going to execute the application.

In this chapter the following topics are discussed:

- Setting up an environment for developing SWT applications
- Widgets and displays
- Several user interface elements (controls), such as labels, texts, lists, buttons, and combos
- Layout styles

Getting started

As previously mentioned, SWT is a widget toolkit for Java. A widget is a GUI element developed to provide interaction with an user. Text, labels, lists and buttons are examples of widgets. For making use of widgets and getting started with the development of SWT applications, some configurations must be performed, as described in the next section.

Setting up

Although SWT is integrated as part of the Eclipse plug-in API, for older versions or for the development of standalone applications, it may be necessary to download the standalone SWT.

For that, one must go to `www.eclipse.org/swt`. After selecting the appropriate version (in our case `swt-4.2.1-gtk-linux-x86.zip`), it is necessary to import the downloaded file into Eclipse by going to **File** | **Import...** | **Existing projects into workspace** | **Select archive file**, and by hitting the **Browse** button to locate the SWT library file, as it is shown in the following screenshots:

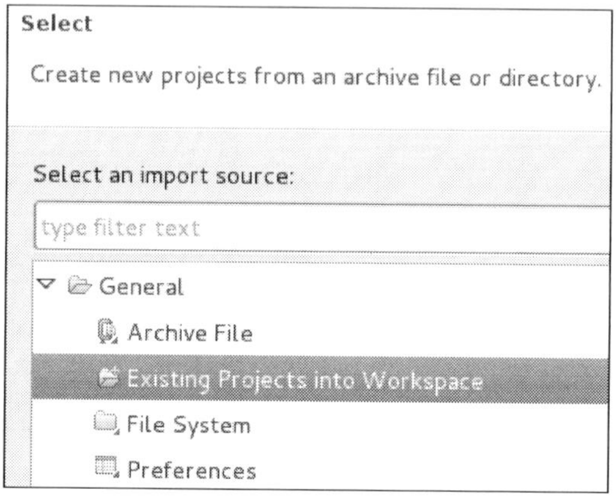

(a) Import Source

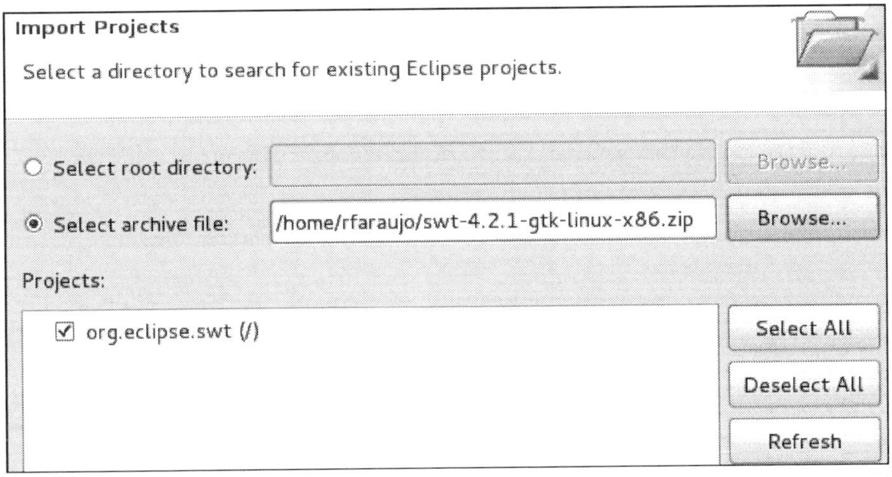

(b) SWT Library Archive File

After these first steps, it is necessary to include the library in the Java build path of the working project. We show the source code of a sample `Hello World` SWT program that is used in our explanation on how to add library in the path, as follows:

```
import org.eclipse.swt.widgets.*;
public class HelloWorld {
   public static void main(String[] args) {
  Display display = new Display();
  Shell shell = new Shell(display);
  shell.setText("Hello, World.");
  shell.open();
  while (!shell.isDisposed()) {
    if (!display.readAndDispatch())
      display.sleep();
  }
  display.dispose();
   }
}
```

In this sample code that results in the output that is shown in the following screenshot, the basics of a SWT application is presented. A connection to the platform window system is created by the `Display` object, in which a window is built using that connection containing the text `Hello, World`. by the `Shell` object. In the loop that follows, the window system waits for an action by the user that would cause the application to terminate.

To properly run and compile this sample SWT program, you will need to add the SWT library project as a dependency. This operation can be performed by opening the **Properties** dialog of your Java project, and on the **Java Build Path** page, including the `org.eclipse.swt` project, as it is shown in the following screenshot:

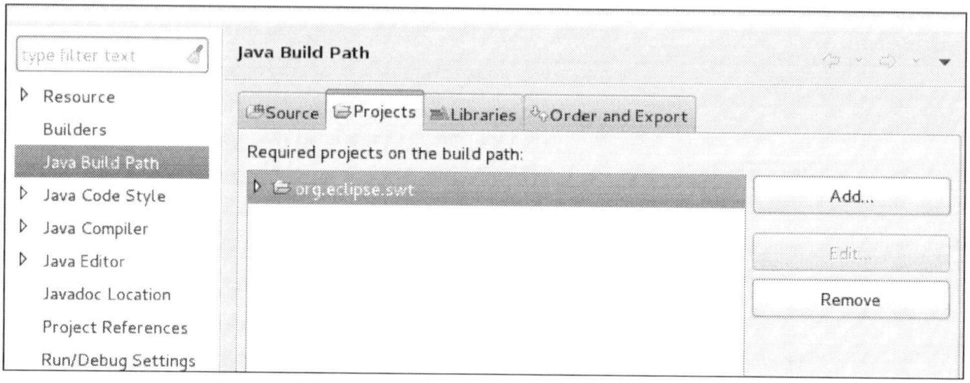

Widgets

Widgets are GUI elements responsible for maintaining the interaction with the user, keeping and redrawing their state, depending on which operation is performed. When an event occurs, which could be the result of an user action or an operation contained inside a program code, a widget redraws itself so that the new state can be shown.

Standard constructors are used to create a widget. As seen in the code snippet of the *Setting up* section , they are explicitly disposed when there is no more need for them. In a different matter from traditional Java classes, you can note this behavior because widgets use operating systems resources that must be manually disposed in order to avoid resource leaks. The Java garbage collector does not dispose operating system resources.

In this context, when an SWT object is created, or when an SWT resource-based object is called, there is a need to manually dispose it. If there the resource has a parent, disposing it also causes the disposal of its children.

A group of widgets instances are used to build a user interface in the SWT. For that we have the control element, that is a user interface contained inside an application window, called a Shell. Several ordinary elements as Buttons, Texts, Labels and Tables subclass the abstract class Control.

To deal with actions, instead of sub-classing to implement code that reacts to modifications in a GUI, programmers are encouraged to use events and listeners that are added to widgets appropriately. Listeners are used to notify the program when an event has happened.

The last main feature that you must be aware of a widget, regards its application data. Due to the discouraged use of sub-classing (due to possible misbehavior regarding modifications in the superclass followed by the non-recompilation of the subclass), application data can be used to associate any object with a widget. The `setData` and `getData` methods can be used for that purpose, as we exemplify as follows:

```
widget.setData("Doctor", "Walter Bishop");
if ("Walter Bishop".equals(widget.getData("Doctor"))) {
  System.out.println("I'm perfectly sane!");
}
```

In this example, we store the data as a key/value pair in the widget, where the key is used to get/find the value object, that is, the `Doctor` key corresponds to the `Walter Bishop` object, and if the `if` statement returns `true`, the `I'm perfectly sane` message is printed.

Real-world situations may require factories to facilitate the storage of data, as it can be seen in the *Widgets* section.

Displays

The `Display` class is responsible for connecting the SWT with the underlying window system. It is not a widget; nevertheless, some of the principles discussed in *Widgets* section are still the same, such as the handling of events, listeners, and application data. The `Display` lifecycle is also similar to that of a widget, and should be explicitly created and disposed by the programmer.

One of the most important methods inside it is the `readAndDispatch` method, that can read and dispatch events. The method reads events from native widgets and dispatches them to the SWT event system. This continuous loop is executed until the main shell is closed. If the loop is interrupted, the application terminates immediately. It is called inside the user-interface thread, where many widgets methods are also called. As seen in the *Setting up* section, standalone applications can create and make calls to the display event processing class inside its `main` method.

It is also possible to name the application via the `setAppName` method, that takes a single `String` as argument.

Controls

A control is an user-interface element that belongs to a higher-level window. In this section, we are going to describe some of the basic ones in detail, such as labels, texts, lists, layouts, buttons, and combos. These simple controls are commonly used when developing applications using the SWT. In the next chapter, we will extend this section with more advanced controls and examples.

The Label widgets

Labels are static widgets that represent an object that cannot be modified or selected, and can act like a string, image, or separator. They are usually placed in the left-hand side of the screen to describe text fields, act as separators among non-related fields or to place an image.

The following styles can be selected when you create a label, and can be combined using the | operator:

- `SWT.SEPARATOR`: Draws a separator
- `SWT.HORIZONTAL`: Draws the separator horizontally
- `SWT.VERTICAL`: Draws the separator vertically
- `SWT.SHADOW_IN`: Separator with shadow in effect

- `SWT.SHADOW_OUT`: Separator with shadow out effect
- `SWT.SHADOW_NONE`: Separator with no shadow effect
- `SWT.CENTER`: Aligns the label to the center
- `SWT.LEFT`: Aligns the label to the left
- `SWT.RIGHT`: Aligns the label to the right
- `SWT.WRAP`: Automatically wraps the text to fit the container

The main methods used when creating a label widget are:

- `setText(String string)`: Sets the text of the label
- `getText()`: Returns the text of the label
- `setImage(Image image)`: Sets an image to be included in the label
- `getImage()`: Returns the image of the label

We show a sample code of how to use a label, along with the widget representing it as follows:

```
Display display = new Display();
Shell shell = new Shell(display);
shell.setText("Label Sample");
shell.setSize(300,200);
Label label = new Label(shell, SWT.CENTER);
label.setText("Nothing But The Rain");
label.setSize(300,300);
```

It bears mentioning that the inherited `setSize` method takes as parameters an horizontal and a vertical size in pixels.

The Text widgets

Text widgets are editable text fields for strings. These strings can comprehend one or more lines, and can be used for password fields. They are also selectable by an user.

The following are several styles available for the creation of a Text widget:

- SEARCH: Searches behavior, can be combined with icon
- ICON_CANCEL: Shows the **Cancel** icon
- ICON_SEARCH: Shows the **Search** icon
- MULTI: Allows multiple lines
- SINGLE: Allows single line
- PASSWORD: Shows the password field behavior
- CENTER: Aligns the text content to the center
- RIGHT: Aligns the text content to the right
- LEFT: Aligns the text content to the left
- READ_ONLY: Makes the field read-only
- WRAP: Wraps the text instead of scroll behavior

Text widgets support only plain text, which means that the characters must have the same font and size. If a more flexible text control is needed, one could use a `StyledText` widget, which is not a native widget from SWT.

When using text fields in an SWT program, it is rather common that the widget will have some kind of functionality in the application. A programmer often needs to detect when a text has changed or to filter a string that is being typed. For this purpose, it may be necessary for listeners to handle the following text-control events:

- `DefaultSelection`: *Enter* is pressed
- `Modify`: Text is modified
- `Verify`: Text is to be validated
- `OrientationChange`: Orientation is changed

In the following example we show how to deal with one of the mentioned events, `SWT.Modify`. For that we create a `Text` widget, and add a listener for the `SWT.Modify` event. This event is sent when any character is entered by an user. The purpose of the code inside such listener is to count the number of characters that have been typed. A message is displayed in a label located underneath the `Text` widget.

```
Display display = new Display();
Shell shell = new Shell(display);
shell.setSize(200,200);
final Text text = new Text(shell, SWT.CENTER | SWT.BORDER);
text.setSize(200,25);
final Label label = new Label(shell, SWT.CENTER);
label.setLocation(0,100);
label.setSize(200,25);
text.addListener(SWT.Modify, new Listener() {
  public void handleEvent(Event event) {
label.setText(Integer.toString(text.getText().length()));
  }
});
shell.pack();
shell.open();
while (!shell.isDisposed()) {
  if (!display.readAndDispatch()) display.sleep();
}
display.dispose();
```

The following screenshot shows the GUI output related to this snippet of code:

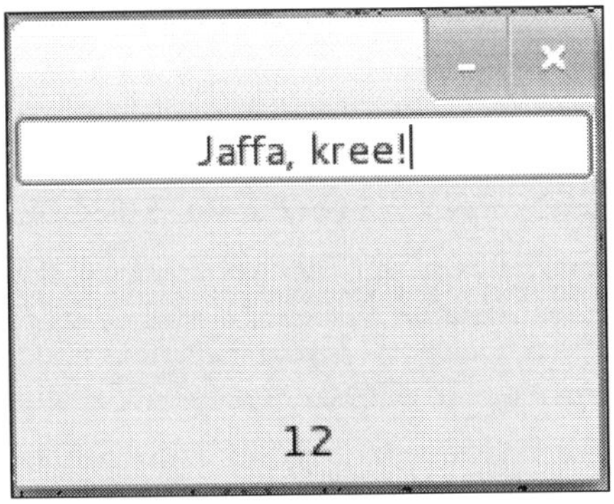

It bears mentioning that in this example we also use the `setSize` and `setLocation` methods, that are respectively responsible for setting the size of the control and its location, by `x` and `y` coordinates in pixels. The `pack()` method is also used to auto resize the widget to its preferred size.

The Button widgets

Buttons are often used in desktop applications because these usually make use of the mouse device. When a user selects an option, it is possible to handle the selection event. Buttons can be filled with text or images, but not at the same time, and there is no option to obtain multiple lines of text on the screen.

The following styles are available when creating a `Button` control:

- `ARROW`: Draws the button with arrow appearance
- `CHECK`: Checks button behavior
- `PUSH`: Pushes button behavior
- `RADIO`: Shows radio button behavior
- `TOGGLE`: Toggles button behavior
- `FLAT`: Draw the button with flat appearance
- `UP`: Draws an up arrow
- `DOWN`: Draws a down arrow
- `LEFT`: Draws a left arrow or aligns to the left
- `RIGHT`: Draws a right arrow or aligns to the right
- `CENTER`: Draws a center arrow or aligns to the center

The only available event is the `SWT.Selection` event, that is triggered when a button is clicked.

We show an example of how to use a `Button` control, followed by the application output as follows:

```
Display display = new Display ();
Shell shell = new Shell (display);
Label label = new Label (shell, SWT.NONE);
label.setText ("Enter your name:");
Text text = new Text (shell, SWT.BORDER);
text.setLayoutData (new RowData (100, SWT.DEFAULT));
final Label resultLabel = new Label(shell, SWT.BORDER);
resultLabel.setText("Nothing pressed");
Button ok = new Button (shell, SWT.PUSH);
ok.setText ("OK");
ok.addListener(SWT.Selection, new Listener() {
  public void handleEvent(Event event) {
    resultLabel.setText("OK pressed");
  }
});
```

```
Button cancel = new Button (shell, SWT.PUSH);
cancel.setText ("Cancel");
cancel.addListener(SWT.Selection, new Listener() {
  public void handleEvent(Event event) {
    resultLabel.setText("Cancel pressed");
  }
});
shell.setLayout (new RowLayout ());
shell.pack ();
shell.open ();
while (!shell.isDisposed ()) {
  if (!display.readAndDispatch ()) display.sleep ();
}
display.dispose ();
```

In this example, we create two push buttons, namely **OK** and **Cancel**, and we add two listeners that are triggered when each button is pushed, showing the message **OK pressed**, or **Cancel pressed**. In a real-world application, a possible use of such application could be reading the name that was typed in the `Text` widget for recording when the **OK** button is pressed.

The List widgets

A list can be used to show a set of selectable strings by the user. They are presented in a single column, and there are methods for selecting, adding, and removing items from it. The contents of a list can be set by the method `setItems(String[] items)`. When you need to set or replace a specific item, it is possible to use `setItem(int index, String item)`.

Similarly, to the `setItem` method, this widget facilitates the achievement of one or more items from a list by using the `getItem` or `getItems` methods.

Items can also be added by the `add` method, and removed using the `remove` method.

The `setSelection` and `getSelection` methods are generally used when operations dealing with the selection of items need to be performed.

The following styles are available for a `List` widget:

- `SINGLE`: Only one item can be selected
- `MULTI`: Multiple items can be selected

The description of each style is pretty straightforward. We present the available events for this control as follows:

- `SWT.Selection`: An item is selected
- `SWT.DefaultSelection`: An item is double-clicked

A code sample using the `SWT.Selection` method is described, followed by the GUI outcome as follows:

```
Display display = new Display ();
Shell shell = new Shell (display);
final List list = new List (shell, SWT.BORDER | SWT.MULTI | SWT.V_
SCROLL);
for (int i=0; i<10; i++) list.add ("Item " + i);
list.setSize(200, 200);
list.addListener (SWT.Selection, new Listener () {
  public void handleEvent (Event e) {
    String string = "";
    int [] selection = list.getSelectionIndices();
    for (int i=0; i<selection.length; i++)
      string += selection[i] + " ";
    System.out.println ("Selected: " + string);
  }
});
shell.pack ();
shell.open ();
while (!shell.isDisposed ()) {
  if (!display.readAndDispatch ()) display.sleep ();
}
display.dispose ();
```

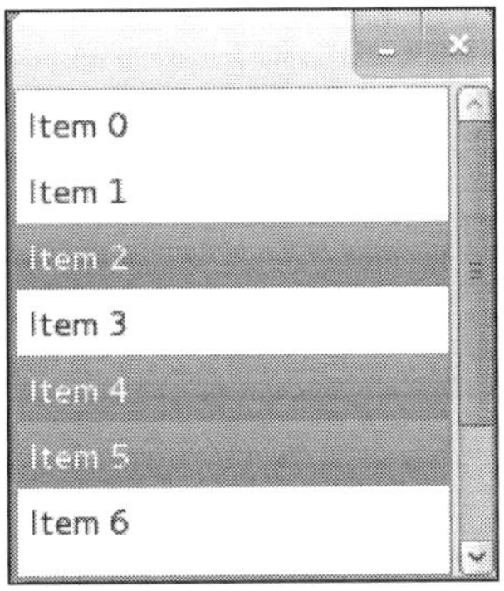

In this example, we added 10 items to a list using the add method, and handled the event `SWT.Selection`. When selecting those three items of our example, the application outputs the following string to the console: **Selected: 2 4 5**.

The Combo widgets

A `Combo` widget, commonly known as combo box, is a combination of a `Text` widget and of a `List` one. The `Combo` widget is generally displayed as a drop-down list when an user clicks on a small icon in the `Combo` control.

It is possible for the user to type in the text field or to choose an item from the list, but if desired, the combo box can be set as read-only, where only the pre-defined item values can be selected.

The methods for the `Combo` widget are a combination of the ones from the `Text` widget together with the ones from the `List` widget.

The following styles are available for the `Combo` widget:

- `SIMPLE`: Combo always has a visible list
- `DROP_DOWN`: Combo is a drop-down list
- `READ_ONLY`: Combo is non-editable

The available events for the `Combo` widget are as follows:

- `DefaultSelection`: Item is double-clicked or *Enter* is pressed
- `Modify`: Text has been modified
- `Selection`: Item is selected

We show an example of how to use a combo box, followed by the GUI output.

```
Display display = new Display ();
Shell shell = new Shell (display);
Combo combo = new Combo (shell, SWT.SIMPLE | SWT.READ_ONLY);
combo.setItems(new String [] {"Three", "Six", "Eight"});
combo.setSize(200, 50);
shell.pack ();
shell.open ();
while (!shell.isDisposed ()) {
  if (!display.readAndDispatch ()) display.sleep ();
}
display.dispose ();
```

In this example, a combo box is created using the simple style merged with the read-only one. Three items are added and no event is being handled.

Layouts

Layouts are classes to facilitate the positioning and sizing of controls inside a top-level window. They are platform independent, and do not share the same behavior as widgets.

The use of layouts is not mandatory when developing a SWT application; nonetheless, their use is highly encouraged. Layouts are a way to facilitate managing and positioning of real-world applications controls, because the use of `setSize` and `setLocation` methods in large programs can be a burden.

Layouts offer some benefits when compared to the positioning by hand. They often have the ability to compute the initial size of controls and to reposition them when a resizing operation occurs.

There are two types of the `layout` classes:

- A layout class that provides a specific API to its implementation
- A layout class that works with a specific layout data algorithm to be assigned for each child control of a composite

A layout data follows an algorithm that makes use of additional information on how the control should be positioned. When using this type of layout, we need to associate the composite with the `layout` class, and the composite children with the layout data.

Regardless, each layout presents a different underlying algorithm, and in the following sections we describe the available SWT layouts.

FillLayout

The simplest layout class is `FillLayout`, which does not work along with a layout data. It positions the controls in a single row or column using the available space, and divides and assigns the same space for each one. If nothing is specified, it assumes that the controls will be as high as the highest one, and as wide as the widest. `FillLayout` is often used when the application has one child, as it will completely fill the shell.

`FillLayout` provides the following public fields:

- `type`: Layout orientation, that can be `SWT.HORIZONTAL` or `SWT.VERTICAL`
- `marginWidth`: Number of pixels of the margin to be placed along the left and right edges of the layout
- `marginHeight`: Number of pixels of the margin to be placed along the top and bottom edges of the layout
- `spacing`: Number of pixels responsible for the space among controls

We show an example of how to position 10 elements of a Text widget as follows. Note that when the application is resized, the controls are also resized by `FillLayout`.

```
Display display = new Display();
Shell shell = new Shell(display);
shell.setLayout(new FillLayout());
for (int i=0; i<10; i++) {
  Text text = new Text(shell, SWT.BORDER);
  text.setText("Fill" + i);
}
shell.pack();
shell.open();
while (!shell.isDisposed()) {
  if (!display.readAndDispatch()) display.sleep();
}
display.dispose();
```

The output of this code snippet is shown as follows:

The following image shows the resize behavior:

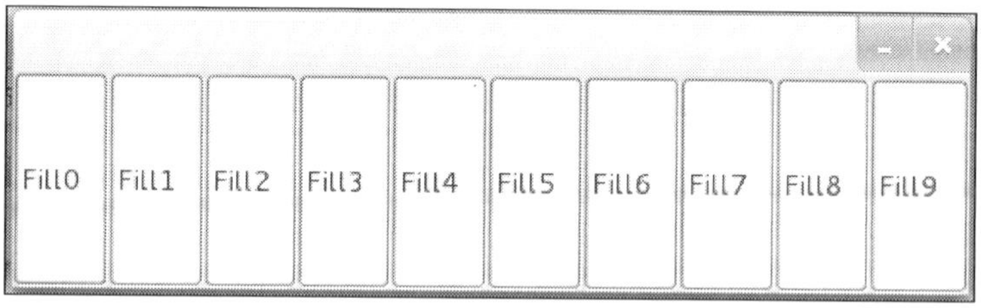

RowLayout

`RowLayout` positions the controls in a row (horizontally) or in a column (vertically), depending on how it was instantiated. It differs from `FillLayout`, as it does not fill out all the available space and it presents a layout data class called `RowData` to configure each of the controls individually. It is possible to specify their height and width when positioning them in the composite.

The following public fields are available:

- `center`: Specifies whether the control should be centered
- `fill`: Specifies whether all controls in a row should be of the same height for horizontal layouts or the same width for vertical layouts
- `justify`: Specifies if any remaining space should be allocated among the controls
- `marginBottom`: Specifies the number of pixels at the bottom edge of the layout
- `marginHeight`: Specifies the number of pixels at the top and bottom edges of the layout
- `marginLeft`: Specifies the number of pixels at the left edge of the layout
- `marginRight`: Specifies the number of pixels at the right edge of the layout
- `marginTop`: Specifies the number of pixels at the top edge of the layout
- `marginWidth`: Specifies the number of pixels at the left and right edges of the layout

- pack: Specifies whether all controls can take their preferred size
- spacing: Specifies the number of pixels among the edges of the cells
- type: Specifies if the layout is using rows or columns
- wrap: Specifies if a control should be wrapped to the next row if there is no space available

RowLayout can be used to position controls both in columns and in rows. Despite its name, if the programmer sets the field to SWT.VERTICAL, the layout orientation will follow this pattern.

In the following example, we show how to position a few buttons using RowLayout and RowData. In the following example, note that in the later two buttons, we use the RowData class to specify the size of the buttons by defining the height and the width:

```
Display display = new Display();
Shell shell = new Shell(display);
RowLayout layout = new RowLayout(SWT.HORIZONTAL);
layout.wrap = true;
layout.fill = false;
layout.justify = true;
shell.setLayout(layout);
Button b = new Button(shell, SWT.PUSH);
b.setText("Button 1");
b = new Button(shell, SWT.PUSH);
b.setText("Button 2");
RowData data = new RowData();
b.setLayoutData(data);
b = new Button(shell, SWT.PUSH);
b.setText("150 vertical");
data = new RowData();
data.height = 150;
b.setLayoutData(data);
b = new Button(shell, SWT.PUSH);
b.setText("150 horizontal");
data = new RowData();
data.width = 150;
b.setLayoutData(data);
shell.pack();
shell.open();
while (!shell.isDisposed()) {
  if (!display.readAndDispatch())
    display.sleep();
}
display.dispose();
```

The output of the sample application is shown as follows:

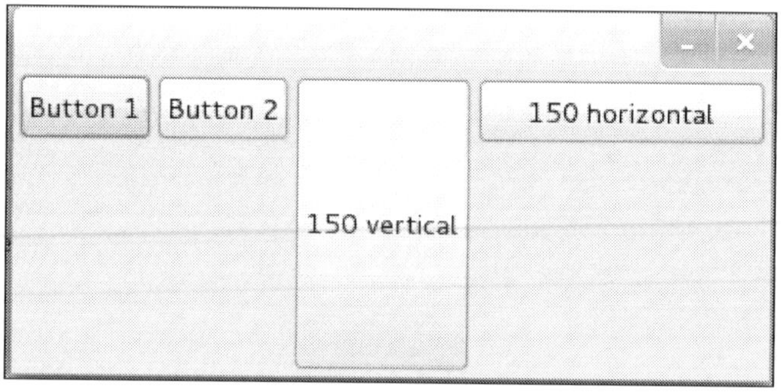

GridLayout

`GridLayout` is an excellent option for dealing with layouts. It has several configuration fields and it is possible to organize the controls of a composite by associating each one with a layout data object provided by the `GridData` class.

In `GridLayout`, it is necessary to specify the number of columns of a grid. The widgets are assigned to the columns in the order they are created, starting from the left and moving to the right. When no more columns are available, the next control is placed in the next row. It provides great flexibility, as each control can also span cells horizontally and vertically.

The following public fields are available for `GridLayout`:

- `horizontalSpacing`: Specifies the number of pixels among the horizontal edges of a cell
- `makeColumnsEqualWidth`: Specifies whether all the columns need to have the same width
- `marginBottom`: Specifies the number of pixels in the vertical margin of the bottom edge of a cell
- `marginHeight`: Specifies the number of pixels in the vertical margin of the bottom and top edges of a cell
- `marginLeft`: Specifies the number of pixels in the horizontal margin of the left edge of a cell
- `marginRight`: Specifies the number of pixels in the horizontal margin of the right edge of a cell
- `marginTop`: Specifies the number of pixels in the vertical margin of the top edge of a cell

- `marginWidth`: Specifies the number of pixels in the horizontal margin of the left and right edges of a cell
- `numColumns`: Specifies the number of cell columns in the `Layout` widget
- `verticalSpacing`: Specifies the number of pixels among the vertical edges of a cell

When applying `GridLayout` to position the elements of an application, it is necessary to create a `GridData` object to provide further information on how this positioning will be performed. We show the main public fields that are available for the `GridData` class:

- `verticalAlignment`: Specifies the vertical alignment
- hortizontalAlignment: Specifies the horizontal alignment
- widthHint: Specifies the amount of space in pixels to be used as the default for a cell width
- heightHint: Specifies the amount of space in pixels to be used as the default for a cell height
- horizontalIndent: Specifies the number of pixels to be placed on the left of the cell
- horizontalSpan: Specifies the number of columns that the cell will fill
- verticalSpan: Specifies the number of rows that the cell will fill
- grabExcessHorizontalSpace: Grabs the horizontal remaining space if it is available
- grabExcessVerticalSpace: Grabs the vertical remaining space if it is available

We show an example of how to use `GridLayout` and `GridData` as follows:

```
Display display = new Display ();
Shell shell = new Shell (display);
GridLayout layout = new GridLayout(4, false);
shell.setLayout(layout);
Button b = new Button(shell, SWT.PUSH);
b.setText("LEFT, TOP");
b.setLayoutData(new GridData(SWT.LEFT, SWT.TOP, true, true, 1, 1));
b = new Button(shell, SWT.PUSH);
b.setText("LEFT, CENTER");
b.setLayoutData(new GridData(SWT.LEFT, SWT.CENTER, true, true, 1, 1));
b = new Button(shell, SWT.PUSH);
b.setText("LEFT, BOTTOM");
b.setLayoutData(new GridData(SWT.LEFT, SWT.BOTTOM, true, true, 1, 1));
b = new Button(shell, SWT.PUSH);
```

```
b.setText("LEFT, FILL");
b.setLayoutData(new GridData(SWT.LEFT, SWT.FILL, true, true, 1, 1));
b = new Button(shell, SWT.PUSH);
b.setText("FILL, TOP");
b.setLayoutData(new GridData(SWT.FILL, SWT.TOP, true, true, 1, 1));
b = new Button(shell, SWT.PUSH);
b.setText("FILL, CENTER");
b.setLayoutData(new GridData(SWT.FILL, SWT.CENTER, true, true, 1, 1));
b = new Button(shell, SWT.PUSH);
b.setText("FILL, BOTTOM");
b.setLayoutData(new GridData(SWT.FILL, SWT.BOTTOM, true, true, 1, 1));
b = new Button(shell, SWT.PUSH);
b.setText("FILL, FILL");
b.setLayoutData(new GridData(SWT.FILL, SWT.FILL, true, true, 1, 1));
shell.open ();
while (!shell.isDisposed ()) {
  if (!display.readAndDispatch ()) display.sleep ();
}
display.dispose ();
```

It is possible to realize in this example that `GridLayout` is receiving two parameters. The first one is the number of columns, and the second one is to specify that the columns do not have the same width. In a similar fashion, the `GridData` method receives six parameters: `horizontalAlignment`, `verticalAlignment`, `grabExcessHorizontalSpace`, `grabExcessVerticalSpace`, `horizontalSpan`, and `verticalSpan`, which were explained previously.

The following GUI output is generated by our example:

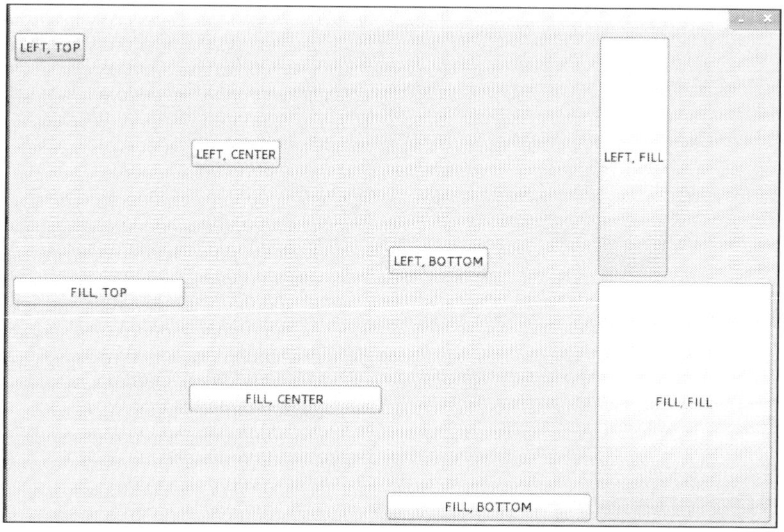

New and noteworthy

In this section, we try to pinpoint some of the novelties regarding the SWT API of the Eclipse Juno version.

The first one regards the `StyledText` widget. When dealing with text that needs customizations, the basic `Text` widget may not be sufficient. In this version, some improvements have been made, such as showing automatic scrollbars only when needed.

Another novelty is the possibility of customizing the default browser renderer. This is particularly significant for programs that make use of browsers with a specific native renderer style. The browser also added support for XULRunner 10.

Some of the other updates that can be mentioned are:

- New methods in the `Combo` API to get caret information
- Improvements in the `TreeCursor` class for navigating among cells
- The `TrayItem` image can be highlighted
- `ColorDialog` added support for custom colors

Summary

In this chapter, you have been introduced to the set of tools known as the SWT, that is designed for the development of graphical interfaces in Eclipse. You have learned some of the basic principles of its widgets, such as texts, buttons, lists, and more. You have also learned how to position this controls using layouts and had a glimpse of how to handle some events. The acquired knowledge is essential for us to understand more complex SWT concepts in the next chapter.

6
More SWT

As previously explained, the **Standard Widget Toolkit** (**SWT**) consists of a set of tools developed by the Eclipse project in order to help building **Graphical User Interface** (**GUI**) applications.

In *Chapter 5, SWT*, we have introduced some basic concepts regarding the SWT API. This knowledge is important so that we can easily follow the content presented in this chapter. For the later sections of this chapter, we delve deep into SWT events, and show how to deal with menus, toolbars, dialogs, and tables.

Events

In several examples of our previous chapter, despite not being the focus of what we were trying to explain, we have shown how to deal with a few user events, such as the click of button, for instance. For every action a user takes, such as a mouse click, or a pressed key, an event is generated by the operating system. We can choose to handle what is considered relevant for the application being developed. To capture an event, it is necessary to add listeners to the code, that must be attached to an interface element, such as a shell, a composite, or a widget, present in a container.

The main process starts with the operating system event queue, that deals (record and list) with actions taken by an user. When an SWT application is being executed, the `Display` class uses the `readAndDispatch` method to monitor the messages from the operating system. It sorts through the relevant ones and sends them to the shell object, which is responsible for redirecting it to the associated widget. The information regarding the event is finally caught by the widget listener, which then invokes a method to handle the event.

In this chapter, we will discuss how to deal with events related to the SWT API, and illustrate our explanation with some examples.

Typed and untyped events

For processing an event, it is necessary to use a listener. In the first versions of the SWT API, there were only untyped listeners. These listeners provide a simple mechanism to handle the events. There is only one generic interface called **Listener** and an event class called **Event**. In order to add a listener, it is necessary to use the `addListener` method.

After some discussion among the Eclipse community, it was decided that a more familiar pattern would be introduced in the SWT API. These are the typed listeners, that were defined in terms of the untyped ones, and they can be found in the `org.eclipse.swt.events` package. We can see the difference among them in the following snippet of code:

```
...
widget.addListener(SWT.Dispose, new Listener() {
    public void handleEvent(Event event) {
        // widget was disposed
    }
});
...
...
widget.addDisposeListener(new DisposeListener() {
    public void widgetDisposed(DisposeEvent event) {
        // widget was disposed
    }
});
...
```

In this example, `DisposeListener` is an interface. When there is more than one method defined in the listener, the SWT API provides an adapter class with these methods. For instance, the `SelectionListener` interface has two methods, `widgetSelected` and `widgetDefaultSelected`, that takes `SelectionEvents` as arguments. As a result, the `SelectionAdapter` class contains both methods.

A table containing the typed events and the corresponding untyped ones is shown as follows:

Event	Listener	Method	Untyped event
`ArmEvent`	`ArmListener`	`widgetArmed`	`SWT.Arm`
`ControlEvent`	`ControlListener`	`controlMoved`	`SWT.Move`
	`ControlAdapter`	`controlResized`	`SWT.Resize`
`DisposeEvent`	`DisposeListener`	`widgetDisposed`	`SWT.Dispose`

Event	Listener	Method	Untyped event
FocusEvent	FocusListener	focusGained	SWT.FocusIn
	FocusAdapter	focusLost	SWT.FocusOut
HelpEvent	HelpListener	helpRequested	SWT.Help
KeyEvent	KeyListener	keyPressed	SWT.KeyPressed
	KeyAdapter	keyReleased	SWT.KeyReleased
MenuEvent	MenuListener	menuHidden	SWT.Hide
	MenuAdapter	menuShown	SWT.Show
ModifyEvent	ModifyListener	modifyText	SWT.Modify
MouseEvent	MouseListener	mouseDoubleClick	SWT. MouseDoubleClick
	MouseAdapter	mouseDown	SWT.MouseDown
		mouseUp	SWT.MouseUp
MouseMoveEvent	MouseMoveListener	MouseMove	SWT.MouseMove
MouseTrackEvent	MouseTrackListener	mouseEnter	SWT.MouseEnter
	MouseTrackAdapter	mouseExit	SWT.MouseExit
		mouseHover	SWT.MouseHover
PaintEvent	PaintListener	paintControl	SWT.Paint
SelectionEvent	SelectionListener	widgetDefaultSelected	SWT. DefaultSelection
	SelectionAdapter	widgetSelected	SWT.Selection
ShellEvent	ShellListener	shellActivated	SWT.Activate
	ShellAdapter	shellClosed	SWT.Close
		shellDeactivated	SWT.Deactivate
		shellDeiconified	SWT.Deiconify
		shellIconified	SWT.Iconify
TraverseEvent	TraverseListener	keyTraversed	SWT.Traverse
TreeEvent	TreeListener	treeCollapsed	SWT.Collapse
	TreeAdapter	treeExpanded	SWT.Expand
VerifyEvent	VerifyListener	verifyText	SWT.Verify

The adapters mentioned in the preceding table are only available for the events whose listeners have more than one method. Adapter objects can be created using a similar structure, the `addXListener`:

```
...
button.addMouseListener(new MouseAdapter()
{
  public void mouseDoubleClick(MouseEvent e)
  {
      System.out.println("Double click.");
  }
)};
  ......
```

Since there are several events and listeners, it is difficult to focus on all of them for an explanation. In the following subsections, we try to give some examples of how to use the most common ones.

KeyEvent

Most of the events are straightforward for understanding and use. Nonetheless, the KeyEvent one, particularly, might need a better explanation. The `KeyEvent` class is created when a key is pressed by the user. It has two subclasses: `TraverseEvent` and `VerifyEvent`. A `TraverseEvent` is generated when the user presses the *arrow* key or the *Tab* key to reach a different widget, and the `VerifyEvent` subclass is activated when the user enters text into a widget.

Three member fields of the `KeyEvent` class typically used are as follows:

- **character**: It provides a char value representing the pressed key
- **stateMask**: It returns an integer that represents the state of the keyboard modification (it is possible to verify if some *Alt*, *Ctrl*, *Shift*, or command keys are being pressed)
- **keyCode**: It provides the SWT public constants corresponding to the typed key

The following snippet of code shows an example of how to use the `KeyEvent` class:

```
public class ComboListenerEx {
        public static void main(String[] args) {
            Display display = new Display();
            Shell s = new Shell(display);

            s.setSize(500,400);
            s.setText("Combo KeyListener");
            s.setLayout(new RowLayout( ));
```

```
        final Combo c = new Combo(s, SWT.DROP_DOWN | SWT.BORDER);
        c.add("Brazil");
        c.add("Argentina");
        c.add("Spain");

        c.addKeyListener(new KeyListener( ) {
            String selectedItem = "";
            public void keyPressed(KeyEvent e) {
                if(c.getText( ).length( ) > 0) {
                    return;
                }
                String key = Character.toString(e.character);
                String[] items = c.getItems( );
                for(int i =0;i<items.length;i++) {
                    if(items[i].toLowerCase( ).startsWith(key.
                        toLowerCase( ))) {
                            c.select(i);
                        selectedItem = items[i];
                        return;
                    }
                }
            }
            public void keyReleased(KeyEvent e) {
                if(selectedItem.length( ) > 0)
                c.setText(selectedItem);
                selectedItem = "";
            }
        });
        s.open( );
        while(!s.isDisposed( )) {
            if(!display.readAndDispatch( ))
                display.sleep( );
        }
        display.dispose( );
    }
}
```

This example creates a simple combo widget. Since it is of the `SWT.DROP_DOWN` style, it is editable. The objective of this piece of code is to select the item corresponding to the first letter pressed by the user. For instance, if the letter pressed was the *B* key, the "Brazil" item would be selected.

MouseEvent

A `MouseEvent` occurs when a mouse button is pressed. A `MouseListener` can be used to handle three types of events:

- `MouseButtonPressed`: This event takes place when a button is pressed, but not released
- `MouseButtonReleased`: This event takes place when a button that has been pressed is released
- `MouseDoubleClick`: This event takes place when a mouse button is double-clicked. Each of the these events have a corresponding method in the listener class that is called when an event takes place.

The following code snippet shows a listener handling mouse events:

```
...
public class MouseExample {
    public static void main(String[] args) {
        Display display = new Display();
        final Shell s = new Shell(display);
        s.setSize(500,400);
        s.setText("MouseEvent");
        s.open();

        s.addMouseListener(new MouseListener() {
                public void mouseDown(MouseEvent e) {
                        System.out.println("Position:" + e.x + " " +
                                                    e.y);
                }
                 public void mouseUp(MouseEvent e) {
                        System.out.println("Position2:" + e.x + " " +
                                                    e.y);
                }
                public void mouseDoubleClick(MouseEvent e) {
                        System.out.println("Double click");
                }
        });
        while(!s.isDisposed( )){
                if(!display.readAndDispatch( ))
                        display.sleep( );
        }
        display.dispose( );
    }
}
...
```

In this example, the `mouseUp` and `mouseDown` methods are triggered when a mouse click occurs. The `mouseDoubleClick` method shows a message in the console when a double-click happens.

Menus

Most graphical application makes use of some sort of menu. Its labels are often referred to as File, Edit, or Exit, for instance, and are usually placed on the top of an application window to display a list of choices. In SWT, the `Menu` and `MenuItem` classes are responsible for handling operations related to menus.

Basically, we can divide menus into three different kinds:

- **Main**: These menus are created using the `SWT.BAR` style below the shell title
- **Drop down**: These menus are also referred to as submenus or cascade menus and are created by using the `SWT.DROP_DOWN` style
- **Pop up**: These menus are created using the `SWT_POP_UP` style and are displayed when an user requests a menu in a control

The relationship among these three types of menus works by having the menu bars and pop-up menus as the root of the hierarchy. Both of them are usually composed of drop-down menus.

Menu acts as a container for `MenuItems`, and contains methods so that `MenuItems` can be added. `MenuItem` is a widget that can be selected by the user or can display another menu. It can be divided into the following categories:

- `SWT.PUSH`: A standard menu item is created in this category
- `SWT.CHECK`: It is used for a checkbox button
- `SWT.RADIO`: It is used for a radio button
- `SWT.SEPARATOR`: It is used to separate several menu items
- `SWT.CASCADE`: It is used to create a submenu in a cascade style

There are several events that can be triggered from a `Menu` object and a `MenuItem` object:

- The `Menu` events are as follows:
    - `SWT.Hide`: Menu is hidden
    - `SWT.Show`: Menu is shown
    - `SWT.Help`: Help for the menu

- The `MenuItem` events are as follows:
    - `SWT.Selection`: Menu item selected
    - `SWT.Arm`: Menu item drawn in armed state
    - `SWT.Help`: Help for the menu item

The following snippet of code shows an example of how to use a `Menu` object and a `MenuItem` object. Their implementation is quite straightforward. Notice that the `MenuItem` object is created using the `Menu` object as basis.

```
    ...
Display display = new Display();
Shell shell = new Shell(display);
Menu menuBar = new Menu(shell, SWT.BAR);
shell.setMenuBar(menuBar);
MenuItem fileItem = new MenuItem(menuBar, SWT.CASCADE);
fileItem.setText("File");
MenuItem editItem = new MenuItem(menuBar, SWT.CASCADE);
editItem.setText("Edit");
Menu fileMenu = new Menu(shell, SWT.DROP_DOWN);
fileItem.setMenu(fileMenu);
String [] fileStrings = {"New", "Exit"};
for (int i=0; i<fileStrings.length; i++) {
      MenuItem item = new MenuItem(fileMenu, SWT.PUSH);
      item.setText(fileStrings [i]);
}
Menu editMenu = new Menu(shell, SWT.DROP_DOWN);
String [] editStrings = {"Copy", "Paste"};
editItem.setMenu(editMenu);
for (int i=0; i<editStrings.length; i++) {
      MenuItem item = new MenuItem(editMenu, SWT.PUSH);
      item.setText(editStrings [i]);
}
shell.pack();
shell.open();
while (!shell.isDisposed()) {
    if (!display.readAndDispatch()) display.sleep();
}
display.dispose();
    ...
```

The following screenshot shows the corresponding graphical output:

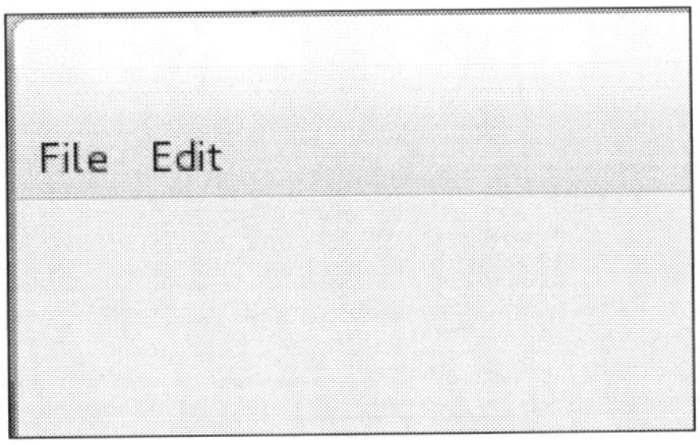

Toolbars

A toolbar is a row of buttons that is shown between the menu and the working area. Despite being rather common, it is not always used by applications. Toolbars can have a text label or an image. Usually, images that provide a hint of the behavior of the buttons are used together with tooltips' help texts when the mouse hovers over the buttons.

The way of creating toolbars is rather similar to creating menus. The `ToolBar` class is considered a container for individual buttons, which are represented by the `ToolItem` objects. The following styles can be used for the `ToolBar` and `ToolItem` objects:

- The `ToolBar` styles are as follows:
    - `SWT.WRAP`: Wrap to fit the visible area
    - `SWT.FLAT`: Toolbar with a flat style
    - `SWT.HORIZONTAL`: Layout buttons horizontally
    - `SWT.VERTICAL`: Layout buttons vertically
    - `SWT.RIGHT`: Place the text in the right of an item
    - `SWT.SHADOW_OUT`: Separator above the toolbar
- The `ToolItem` styles are as follows:
    - `SWT.PUSH`: Create a push button
    - `SWT.CHECK`: Create a check button
    - `SWT.RADIO`: Create a radio button
    - `SWT.SEPARATOR`: Create a separator
    - `SWT.DROP_DOWN`: Create a drop-down item

- The `ToolItem` events are as follows:
    - `SWT.Selection`: Tool item was selected

In the following code snippet, we show a very simple example regarding the creation of a toolbar along with tool items. It is also possible to use images instead of texts.

```
...
Display display = new Display();
Shell shell = new Shell(display);
ToolBar toolBar = new ToolBar(shell, SWT.HORIZONTAL);

ToolItem item = new ToolItem(toolBar, SWT.DROP_DOWN);
item.setText("Drop Down Behavior (Combo)");
toolBar.pack();
shell.pack();
shell.open();
while (!shell.isDisposed()) {
    if (!display.readAndDispatch()) display.sleep();
}
display.dispose();

...
```

We can see that a `ToolBar` item is created, and afterwards a tool item with a drop-down behavior is inserted into it. In a real-world application, it would probably be necessary to add a listener to the `ToolItems` objects, in order that the generated events could be handled.

Tables

A table can be considered a two-dimensional grid composed of several cells. A possible use for tables is the storage of the results of database queries. In order to add data to a table, it is necessary to use a `TableItem` object. Each instance of a `TableItem` object is responsible for an entire row of a table. The values can be set using the `setText` and `setImage` methods that takes an integer as parameter to identify which column should be modified.

In a similar fashion to the `TableItem` object, the `TableColumn` object receives a table as a parameter in order to associate both objects. Each time a `TableColumn` object is associated, a new column is created in the corresponding table.

The following code snippet shows how to create a SWT table:

```
...
 Display display = new Display();
 Shell shell = new Shell(display);

 shell.setSize(500,400);
 shell.setText("Table Example");
 shell.setLayout(new FillLayout( ));

 Table table = new Table(shell, SWT.BORDER);
 TableColumn tc1 = new TableColumn(table, SWT.CENTER);
 TableColumn tc2 = new TableColumn(table, SWT.CENTER);
 TableColumn tc3 = new TableColumn(table, SWT.CENTER);
 tc1.setText("Name");
 tc2.setText("Age");
 tc3.setText("Country");
 tc1.setWidth(80);
 tc2.setWidth(40);
 tc3.setWidth(80);
 table.setHeaderVisible(true);
 TableItem item1 = new TableItem(table,SWT.NONE);
 item1.setText(new String[] {"Rodrigo","28","France"});
 TableItem item2 = new TableItem(table,SWT.NONE);
 item2.setText(new String[] {"Vinicius","28","USA"});
 TableItem item3 = new TableItem(table,SWT.NONE);
 item3.setText(new String[] {"Rafael","28","Brazil"});
 shell.open( );
 while(!shell.isDisposed()){
     if(!display.readAndDispatch( ))
         display.sleep( );
 }
 display.dispose();
...
```

The table GUI regarding the corresponding code is presented as follows:

Despite being possible to set attributes directly on a column, it is a good practice to make use of the `TableLayout` class. `TableLayout` provides the `addColumnData` method, which makes it easier to describe the appearance of each column in the table.

JFace is a UI toolkit with classes for handling common UI programming tasks. JFace is designed to work with SWT and is system dependent. Its most relevant features are actions and viewers. The **action** mechanism allows user commands to be defined independently from their exact whereabouts in the UI. **Viewers** are model-based adapters for certain SWT widgets, simplifying the presentation of application data structured as lists, tables or trees, which is used in our example. `TableLayout` is part of the `org.eclipse.jface.viewers` package. It can be imported in a similar fashion to the SWT, which is explained in the beginning of the previous chapter by going to the following website: `http://wiki.eclipse.org/JFace`

The following code snippet shows how to use a `TableLayout` object:

```
...
TableLayout layout = new TableLayout();
layout.addColumnData(new ColumnWeightData(40, 80, true));
layout.addColumnData(new ColumnWeightData(40, 80, true));
layout.addColumnData(new ColumnWeightData(40, 80, true));
Table table = new Table(shell, SWT.SINGLE);
table.setLayout(layout);
TableColumn column1 = new TableColumn(table, SWT.CENTER);
TableColumn column2 = new TableColumn(table, SWT.CENTER);
TableColumn column3 = new TableColumn(table, SWT.CENTER);
TableItem item = new TableItem(table, SWT.NONE);
```

```
item.setText( new String[] { "column 1",
                              "column 2",
                              "column 3" } );
item = new TableItem(table, SWT.NONE);
item.setText( new String[] { "a", "b", "c" } );
...
```

Each time `addColumnData` is created, it adds a new column to the table. The `ColumnWeightData` object takes as parameter the weight, minimum width, and whether it is resizeable or not. It is important to remember that it is necessary to instantiate the `TableLayout` object to handle the column characteristics and also the `TableColumns` themselves.

Dialogs

Dialog boxes are windows that are separated from the main application. It is usually placed in front of the user to catch his or her attention to an operation that needs to be performed. The dialog abstract class provides the basis for dialogs in SWT. It is possible to extend it in order to create a custom dialog or make use of the several templates already available, such as a confirmation message box or a file dialog.

MessageBox

The `MessageBox` dialog is typically used to display information to the user, such as a possible error, or to obtain input from the user on an operation that needs to be performed. The dialog can be displayed with several buttons, such as yes or no. When the user makes a selection, the corresponding code branch is executed.

The trivial thing to do when using the `MessageBox` dialog is to set its message through the `setMessage` method. Nonetheless, there is a variety of icon styles along with several types of buttons that can be specified. We show the possible styles combinations, and afterwards, the description of the available icons, as follows:

- The styles combinations are as follows:
    - `SWT.OK`
    - `SWT.OK | SWT.CANCEL`
    - `SWT.YES | SWT.NO`
    - `SWT.YES | SWT.NO | SWT.CANCEL`
    - `SWT.RETRY | SWT.CANCEL`
    - `SWT.ABORT | SWT.RETRY | SWT.IGNORE`

- The description of the available dialog icon are as follows:
    - `SWT.ERROR_ICON`: It is used to inform that an error has occurred
    - `SWT.ICON_INFORMATION`: It is used to give non-interactive information
    - `SWT.ICON_QUESTION`: It is used for choosing an answer for the given question
    - `SWT.ICON_WARNING`: It is used to inform the user about a potentially harmful action
    - `SWT.ICON_WORKING`: It tells the user that the work is in progress

The following code snippet shows the use of the `MessageBox` dialog:

```
...
MessageBox messageBox = new MessageBox(shell, SWT.ICON_QUESTION | SWT.
YES | SWT.NO);
messageBox.setMessage("Do you really want to quit?");
messageBox.setText("Exiting Application");
int response = messageBox.open( );
if (response==SWT.YES) {
    shell.dispose();
      System.exit(0);
}
...
```

The following screenshot shows the result of the `MessageBox` example:

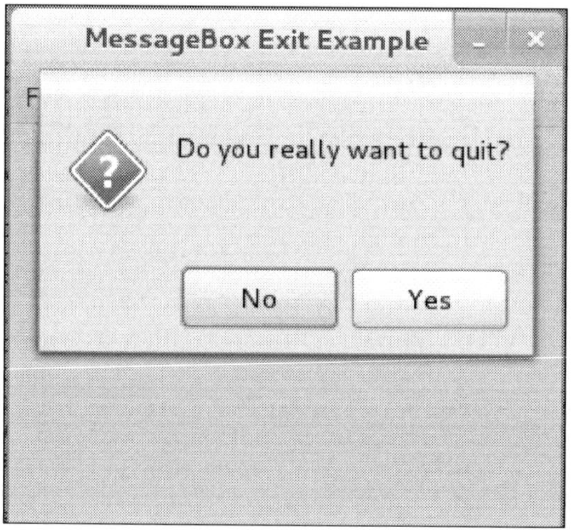

FileDialog

The `FileDialog` opens a dialog with buttons and a list box or tree view that enables the navigation through the filesystem. The programmer only needs to instantiate this class, specifying a few parameters, and does not need to perform any heavy programming.

```
...
FileDialog fd = new FileDialog(shell, SWT.OPEN);
fd.setText("Open");
fd.setFilterPath("/");
String[] filterExt = {"*.txt"};
fd.setFilterExtensions(filterExt);
String selected = fd.open( );
System.out.println(selected);
...
```

The corresponding GUI dialog is shown in the following screenshot:

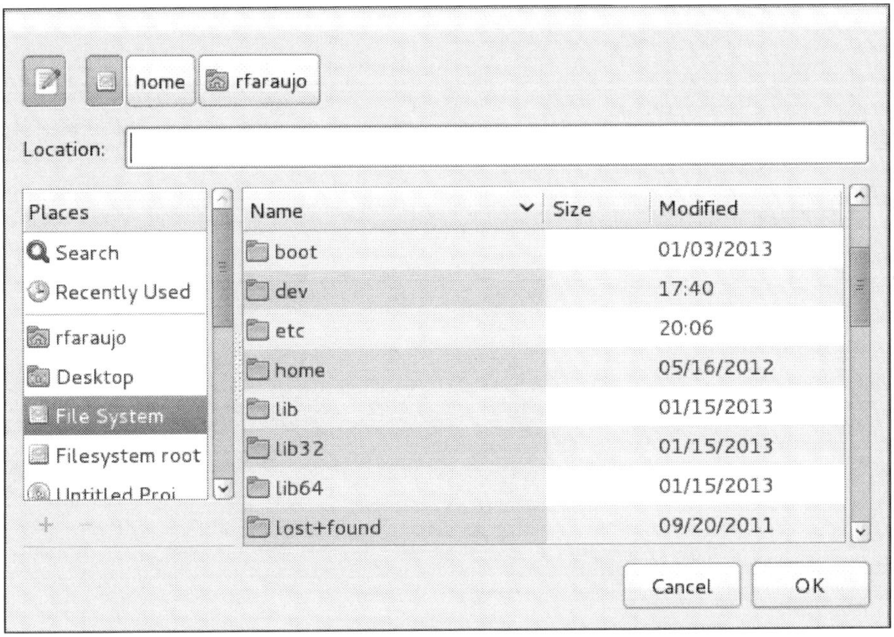

ColorDialog

The `ColorDialog` has the objective of allowing the selection of a color in a widget or text. It is possible to select the desired color graphically, which can then be passed to a widget. It is also possible to select the RGB values manually.

```
...
Text text = new Text(shell, SWT.CENTER | SWT.BORDER);
text.setSize(200,25);
ColorDialog cd = new ColorDialog(shell);
cd.setText("ColorDialog");
cd.setRGB(new RGB(255,255,255));
RGB newColor = cd.open( );
text.setBackground(new Color(display, newColor));
...
```

The corresponding widget can be seen as follows. Note that this widget is dependent on the operating system and may vary among them.

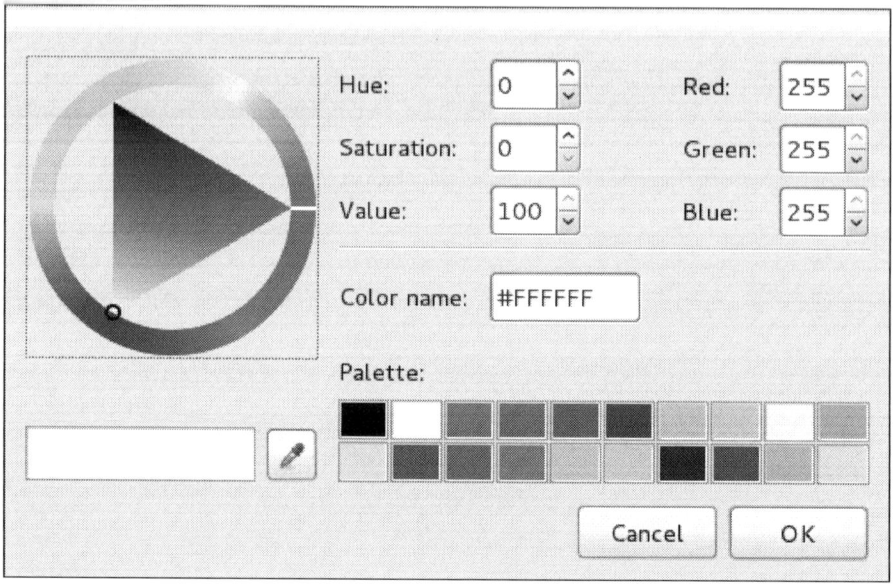

FontDialog

The `FontDialog` can be used for selecting fonts for a text widget in a graphical user interface. Its use is similar to the aforementioned examples.

A snippet of code along with the corresponding output is shown as follows:

```
...
Text text = new Text(shell, SWT.CENTER | SWT.BORDER);
FontDialog fd = new FontDialog(shell, SWT.NONE);
fd.setText("Font Selection");
fd.setRGB(new RGB(0,0,255));
FontData defaultFont = new FontData("Verdana",10,SWT.BOLD);
fd.setFontData(defaultFont);
FontData newFont = fd.open( );
text.setFont(new Font(display, newFont));
text.setForeground(new Color(display, fd.getRGB()));
...
```

The corresponding output is as follows:

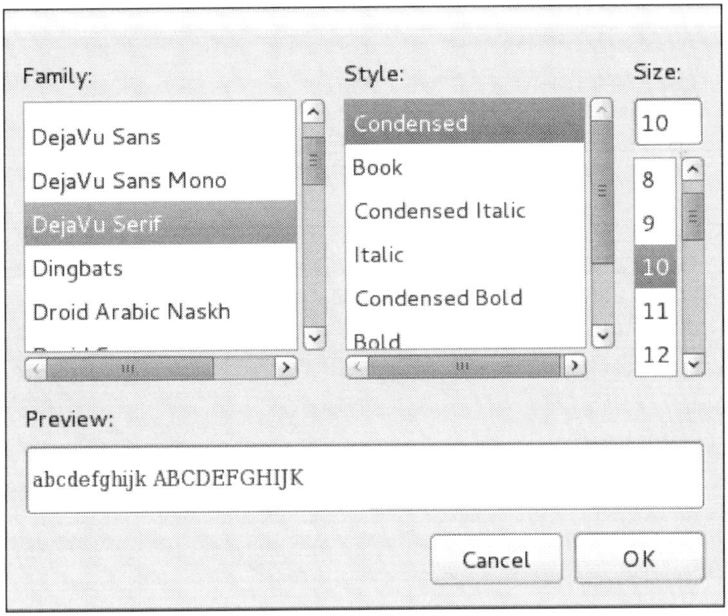

PrintDialog

The `PrintDialog` allows the user to select a printer device to which an application can direct its output. In order to use it, the programmer must follow the same process as previously explained. We present a snippet of code regarding the use of PrintDialog:

```
...
PrintDialog printDialog = new PrintDialog(shell, SWT.NONE);
printDialog.setText("Print");
PrinterData printerData = printDialog.open();
Printer p = new Printer(printerData);
p.startJob("PrintJob");
p.startPage();
Rectangle trim = p.computeTrim(0, 0, 0, 0);
Point dpi = p.getDPI();
int leftMargin = dpi.x + trim.x;
int topMargin = dpi.y / 2 + trim.y;
GC gc = new GC(p);
Font font = gc.getFont();
String printText = "printExample";
Point extent = gc.stringExtent(printText);
gc.drawString(printText, leftMargin, topMargin + font.getFontData()
[0].getHeight());
p.endPage();
p.endJob();
p.dispose();
...
```

If this code is executed, the system configuration printer dialog is presented to the user, and it will print a simple sentence. In this example, the margins are set and a string is drawn using a specified font.

Summary

In this chapter, you have been introduced to more advanced tools of the SWT API that are designed for the development of graphical interfaces in Eclipse. You have learned some of its principles, such as toolbars, tables, menus, and dialogs. We have also focused on how to deal with the SWT events that can be divided into typed and untyped events.

In the next chapter, we will discuss one of the latest trending technologies, the web development, by using Eclipse and Apache Tomcat as a basis.

7
Web Development Using Eclipse WTP

In this chapter, we are going to take a break from using Eclipse to develop applications for the client side. Instead, we are going to explore how Eclipse can be used to develop web-based Java applications. In order to develop such applications, we are going to use Java servlets and **JavaServer Pages** (**JSP**). To run and deploy our example applications, we are going to use Apache Tomcat. Different from desktop applications that run in a standalone fashion, web applications run in a server. Thus, this chapter also details how to deploy Java servlets and JavaServer Pages on a server, namely, Apache Tomcat.

This chapter starts off with a brief introduction to the benefits of Java-based web technologies. Then, we present some background on servlets and JavaServer Pages. We also briefly describe the role that Apache Tomcat plays in implementing these technologies. If you are already familiar with these topics, just skim these first sections and skip to the next subject: installing Apache Tomcat. You will also learn how to integrate Tomcat into Eclipse IDE so that you do not need to leave your development environment to deploy and run your web applications.

We will cover the following in this chapter:

- Introduction to the benefits of the Java-based web technologies
- Servlets
- JavaServer Pages
- Installing Eclipse web tools
- Developing and deploying your first servlet
- Developing and deploying your first JavaServer Page

Brief introduction to the benefits of Java web application technologies

Over the years, a multitude of ways of writing server-side software has been explored. In the early days, the **Common Gateway Interface** (**CGI**) was the most widely used technology to develop web content. However, CGI has many shortcomings. Because of these drawbacks, several technologies have been proposed to replace CGI. Java-based technologies, in particular servlets and JavaServer Pages, rapidly became de facto standard in building web-based applications. Java's widespread adoption in the server side can be ascribed to several factors. First, Java is usually faster than scripting languages. This is especially true for long-running applications because nowadays mainstream Java virtual machine implementations can perform adaptive optimizations that yield substantial performance improvements. Second, older technologies, such as CGI, require the server to spawn a separate process to deal with each request, whereas servlets can be executed in a multithreaded fashion. Finally, Java bytecodes are platform neutral, which allows for applications to be ported to different platforms without significant changes.

A number of enhancements to CGI have been proposed, such as FastCGI, mod_perl from Apache, and NSAPI from Netscape. These technologies solve the performance issue that hindered CGI from becoming even more widespread adopted. However, all of these alternatives to CGI share the same problem: HTML code is embedded directly into programming language code. Thus, the developer writing web applications in these technologies needs to have a mixed skill set: programming language skills for writing the business logic and web-related HTML skills for presentation. This is not desirable for a number of reasons. While servlets suffer from the same problem, JavaServer Pages overcomes this problem by making it possible to separate business logic and presentation properly, as we describe in our quick overview in the following sections.

> You might be wondering what is HTTP and what does HTTP stand for. HTTP is a network protocol, it can be seen as the mean through which clients and servers communicate. The conversation between these two elements usually takes the following sequence: a client (browser) sends an HTTP request, which is answered with an HTTP response by the server. Usually, this response is an **HTML** (**HyperText Markup Language**) page, which is sent using HTTP. HTML tells the client how to show the server's response: the client reads through the message, renders it into a web page and shows it in the user's display. **HTTP** stands for **HyperText Transfer Protocol**.

Understanding servlets

Servlets are Java programs that run within a web server environment. More specifically, servlets are Java programs that conform to the servlet API. Two packages make up most of the functionalities of the servlet API, namely, `javax.servlet` and `javax.servlet.http`. The former encompasses the most general servlet APIs. The latter contains the API for handling HTTP requests. `javax.servlet.http` has a number of classes and interfaces whose main goal is to implement HTTP functionality. One of the most important classes in the package `javax.servlet.http` is the abstract class `HttpServlet`. Servlets inherit their HTTP-related methods from this class. Custom-developed servlets override one or more methods defined in `HttpServlet`. These methods receive an object as a parameter that contains data about the underlying request and an object that they use to write the response. We will go into details about these methods and their parameters later on in the chapter.

> It is common to write servlets that are HTTP-related. In principle, however, it is possible to implement servlets that are geared towards other client-server protocols. Throughout this chapter we are only going to implement HTTP-based servlets.

Understanding JavaServer Page

Put simply, JavaServer Pages are technology geared towards facilitating the development of web pages that contain dynamic content. Differently from an HTML page whose content stays pretty much the same, a JSP changes its content based on a number of things as, for instance, the information previously provided by the user. JSPs consist of HTML or **XML** (**eXtensible Markup Language**) content embedded with tags and Java scriptlets. JSPs can be seen as a more convenient approach for embedding dynamic elements into HTML than servlets. In fact, at run time, JSPs are translated into servlets. Moreover, aimed at increasing performance, servlet containers cache the compiled servlets.

Getting started with Apache Tomcat

Servlets are deployed into a servlet container, which is the component that runs and keeps track of several aspects related to servlets. An example of a servlet container is Apache Tomcat (or simply Tomcat). Put succinctly, Tomcat is an open source web server and Java servlet container from the Apache Software Foundation (`http://tomcat.apache.org`). Tomcat is written in Java, and it implements both the Java servlet and the JSP specifications from Oracle Corporations. Being primarily a servlet container, Tomcat plays a central role in the lifecycle of servlets: it loads classes, instantiates and initializes servlets, maps URLs to servlets, enforces the correct access rights to servlets, and creates (or allocates) threads to run servlets. Roughly, Tomcat is to servlets and JSPs what the JVM is to desktop-based applications— nevertheless, keep in mind that Tomcat runs on the JVM.

Tomcat has a plethora of features, configurations, and options. It is possible to get by with knowing only a small part of all that (which we try to cover throughout this chapter), but you should know that there is a lot more to Tomcat than just downloading it and getting it up and running. However, an in-depth explanation of this full-fledged servlet container is out of the scope of this book. If you are looking for a detailed reference on how to get up to speed on using Tomcat, be sure to read Apache Tomcat 7 Essentials by Tanuj Khare.

Installing Tomcat

Before integrating Tomcat with Eclipse, you need to download and install Tomcat. There are several ways to get Tomcat up and running. On one end of the spectrum, it is possible to download Tomcat's source code and build and test it. On the other end, you can simply download a binary version of it. Here, we are taking the arguably easiest path to installing Tomcat: downloading, configuring, and running the compiled binary. Throughout this chapter, we assume you are using the most recent release of Tomcat. As of the time of this writing, the latest version is 7.0.

> It is also possible to install Tomcat by getting Eclipse to download it for you. Provided that you already have WTP installed (how to install this plugin is described later on), when you create a new server under Eclipse and set its type as **Apache Tomcat**, you can use the **Download and Install** button on the server definition wizard to install Tomcat. You can even select in which folder you want to install the server container.

You can download the binary version of the latest version at `http://tomcat.apache.org/download-70.cgi`. After being downloaded, Tomcat can be installed by simply unzipping it, which creates the following directory hierarchy:

```
Directory Hierarchy
Tomcat
  |
  |___ bin
  |
  |___ conf
  |
  |___ lib
  |
  |___ logs
  |
  |___ temp
  |
  |___ webapps
  |
  |___ work
```

Arguably, the most important directory is the `webapps` directory. This is the deployment directory, and files that need to be made available to client browsers, JAR files, configuration files, and so on are placed into it. Throughout this chapter we are going to provide details on how to organize web applications under this directory.

To set up and run Tomcat outside Eclipse, you need to configure two environment variables, namely, `JAVA_HOME` and `CATALINA_HOME`. The `JAVA_HOME` environment variable should be set to the base path of your JDK. Tomcat 7.0 requires Java SE 6.0 or later. `CATALINA_HOME` should point to the base path of the Tomcat directory hierarchy. Instructions on how to set these environment variables for all operating systems can be found at `http://tomcat.apache.org/tomcat-7.0-doc/setup.html`. In addition, the file `RUNNING.txt` in the Tomcat's root directory also contains installation instructions.

> It is worth mentioning that, although we use Tomcat throughout this chapter, the examples shown here are not Tomcat-specific. Therefore, you can deploy and run all examples shown here in other servlet containers.

At first we are going to execute Tomcat outside Eclipse, later in this chapter we are going to explain how to execute it inside Eclipse. After setting the aforementioned environment variables (following the instructions for your operating system, as described in the URL provided previously), you need to change to Tomcat's `bin` directory. From there, you just need to type `startup` if you are a Windows user or `startup.sh` if you are a Unix-based operating system user. If everything went well, Tomcat has started. To verify if Tomcat is up and running, open a browser and type `http://localhost:8080`. You should see Tomcat's welcome page, as shown in the following screenshot. A few words about this URL are in order. First, `localhost` is what you use to access local web servers. This part of the URL resolves to the following IP address: 127.0.0.1. As for the `8080` part, this corresponds to the port number. Usually, web servers use port 80. Tomcat, however, is configured to listen on port 8080 to avoid possible conflicts with other web servers.

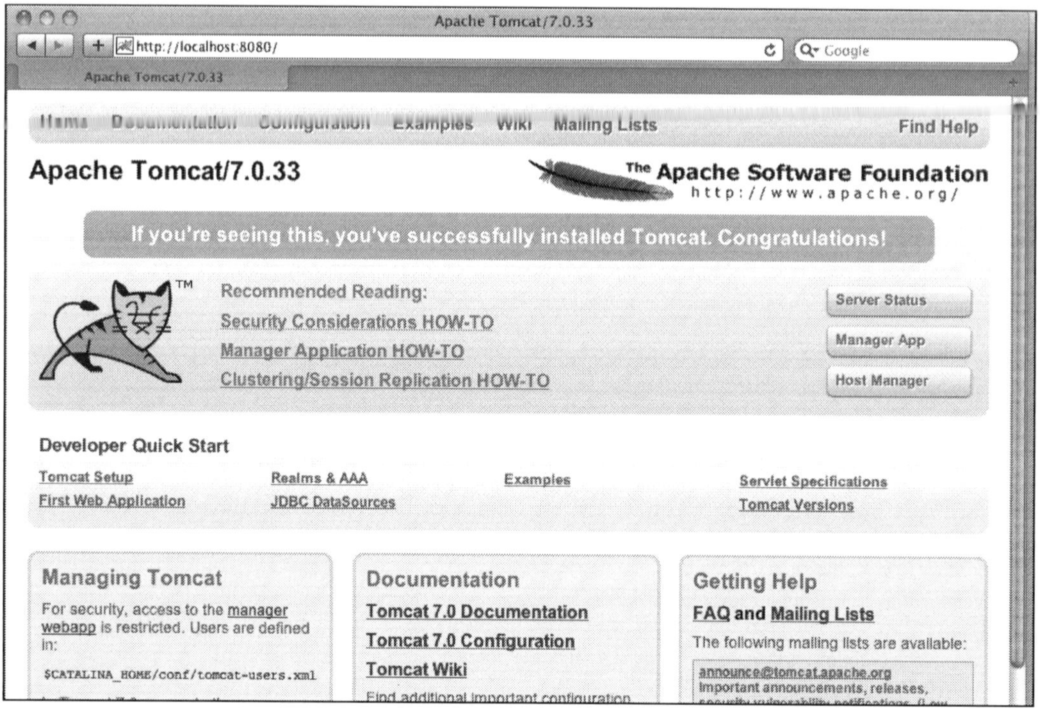

> In some operating systems, for instance, Ubuntu, Tomcat is started automatically. Thus, whenever you want to stop and start Tomcat, use the scripts `stop` and `restart` in the `bin` directory.

Tomcat ships with a web-based administration console. This application is named Manager Webapp, and it starts automatically along with Tomcat. However, you need to do some configuring before being able to use it. In order to be allowed to log in and use the Manager Webapp, you must first configure the file `tomcat-users.xml`, which is in the `conf` directory. Edit this file by adding the following lines:

```
<tomcat-users>
   <role rolename="manager-gui"/>
   <user username="tomcat" password="tomcat" roles="manager-gui"/>
</tomcat-users>
```

Remember that after modifying Tomcat's configuration files, the changes do not take place right away. To indicate Tomcat that something has changed you need to stop the server and then restart it. Upon restarting Tomcat, go to the following URL to access the application:

`http://localhost:8080/manager/html`

The HTML interface of the Manager Webapp is shown in the following screenshot. As you can see, it lists all deployed web applications. The Manager allows you to install, reload, undeploy, and stop web applications. If you want to learn how to perform these operations using the Manager Webapp, then check the accompanying documentation.

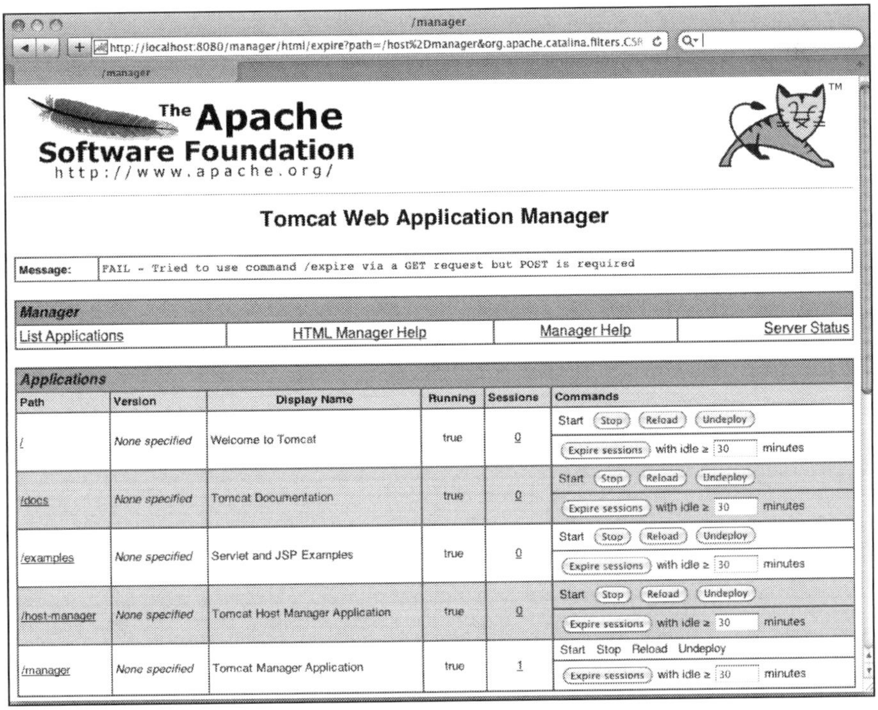

Eclipse Web Tools Platform

We are going to use an extension of the Eclipse platform to help us write, deploy, and run this chapter's applications. To be more specific, we are going to use the Eclipse **Web Tools Platform (WTP)**, which is a set of plugins containing tools for supporting the development and deployment of Java web-based applications. Among other things, using WTP, you can start, stop, and restart the web server from within Eclipse. Thus, you do not need to switch back and forth between Eclipse and Tomcat (or the server of your choice) as you deploy web applications. For an overview of all Eclipse WTP's plugins and their features, visit the project's site: `http://www.eclipse.org/webtools/`.

You can download Eclipse WTP as a binary distribution from the project site: `http://download.eclipse.org/webtools/downloads/`. Nevertheless, this is not the best approach. The most straightforward approach to get WTP up and running is to download the Eclipse version that is prepackaged with the WTP components: Eclipse IDE for Java EE Developers on the `www.eclipse.org/downloads` page. However, most Eclipse versions do not ship with Eclipse WTP. As of this writing, only Eclipse 4.2 has a version that ships with WTP. If you need to use Eclipse 3.8 with WTP, you will have to use the method explained in this chapter.

Alternatively, if you downloaded the Eclipse version for Java development, you can install Eclipse WTP by updating your IDE via the Eclipse Update Manager, which is also an easy approach.

In order to use the Eclipse Update Manager to install Eclipse WTP, click on **Help**, and then on **Install New Software**. You need to type the location of a software site or select an existing one using the **Work with** combo at the top of the installation window. Go ahead and type the following in the **Work with** field:

`http://download.eclipse.org/releases/juno/` (as shown in the following screenshot):

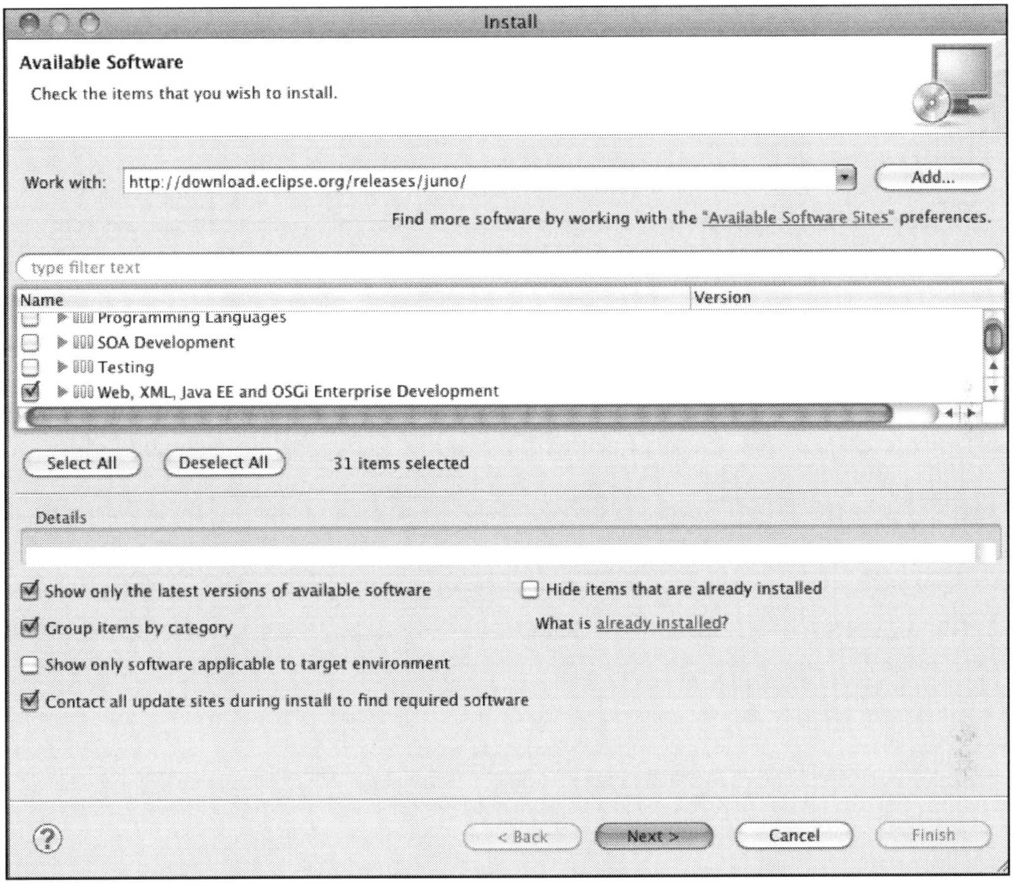

The preceding screenshot shows a list of available updates. As you browse the available software options, select the item **Web, XML, Java EE and OSGi Enterprise Development**, as shown in the previous screenshot. Make sure to check all items belonging to the **Web, XML, Java EE and OSGi Enterprise Development** item but **PHP Development Tools (PDT) SDK Feature**. After selecting the desired items, click on **Next** to install them. The following is what happens when you click on **Next**:

- Once you hit **Next**, Eclipse validates your selections against your installed software. Also, Eclipse takes care of managing dependencies between the selected plugins. Thus, if a selected item requires another software item to operate, the required software will also be downloaded automatically. That is one of the reasons that using Eclipse Update Manager is an easier approach.
- After you agree with the license agreement (select the **I accept the terms of the license agreements** radio button), Eclipse starts installing the chosen extensions.
- After downloading everything, Eclipse will prompt you to restart; click on **Yes** to restart Eclipse.

If everything turns out well, you will be greeted with the following screenshot (something similar to what you saw when you opened Eclipse for the first time):

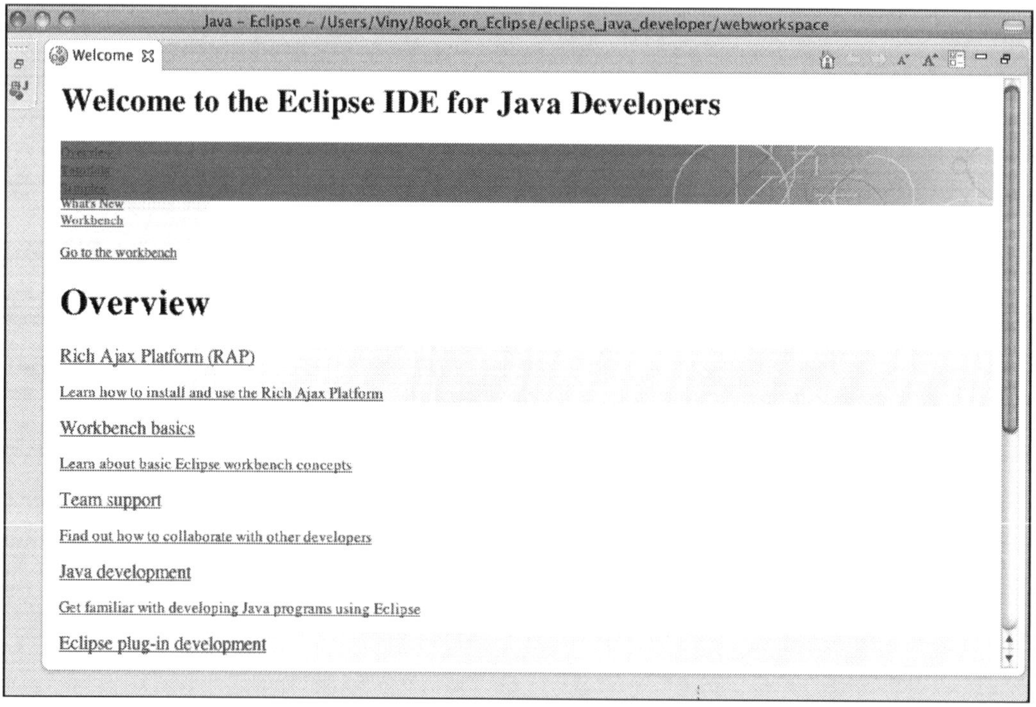

Configuring the server

Now, you have successfully installed Eclipse WTP. However, there is still some configuration to be done before jumping headlong into developing web applications using Eclipse. The next step is setting Eclipse WTP up by indicating which server you want to use and where it is installed (that is, where you unzipped Tomcat). Let's begin by telling Eclipse which server we want to use. To do so, you need to navigate to **Preferences** | **Server** | **Runtime Environments**. You will see the **Preferences** window as shown in the following screenshot. During development, you will be back to this window whenever you feel the need to add or remove servers:

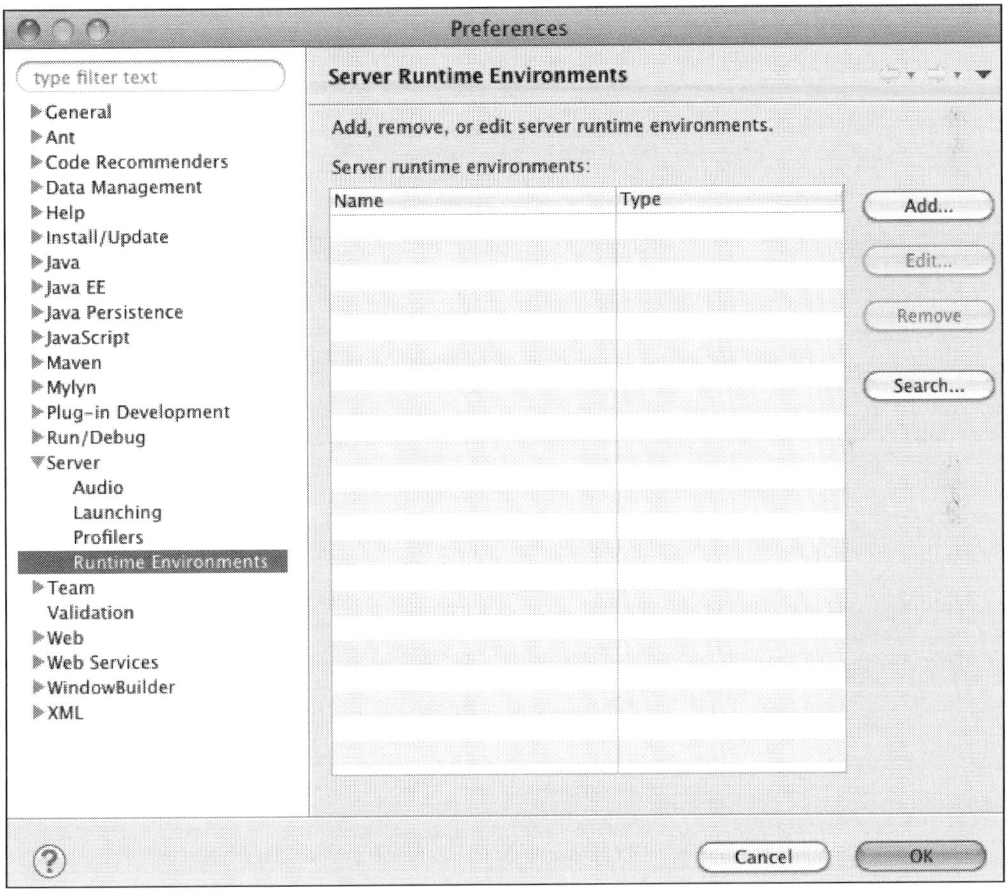

Next, press the **Add...** button on the right-hand side of the dialogue box. A window containing a list of servers will pop up. Apart from Tomcat, Eclipse WTP supports a number of other servers. In case you do not find the desired server in the list, you can download additional server adapters (as shown in the top-right corner of the window in the following figure). To select Tomcat, expand the `Apache` folder, and then pick your desired version of Tomcat. Throughout this chapter we are going to use Apache Tomcat v7.0 (as shown in the following screenshot). After selecting the proper version of Tomcat, you should hit **Next**:

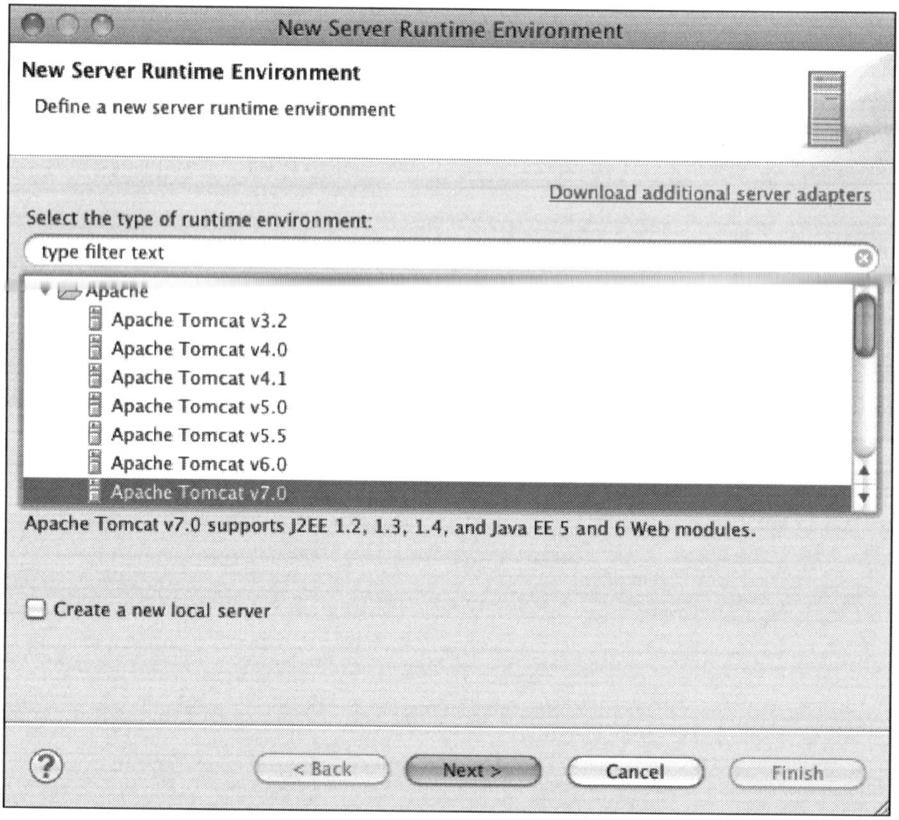

Apart from selecting the desired server runtime environment, you also need to indicate its installation directory. As shown in the following screenshot, in order to inform Eclipse where Tomcat is installed, you need to type the root directory where you unzipped Tomcat. In fact, you can browse to and select your Tomcat root directory by clicking on **Browse...**. For instance, in the following screenshot, the Tomcat root directory is `/usr/local/apache-tomcat-7.0.33`. Afterwards, click on **Finish**.

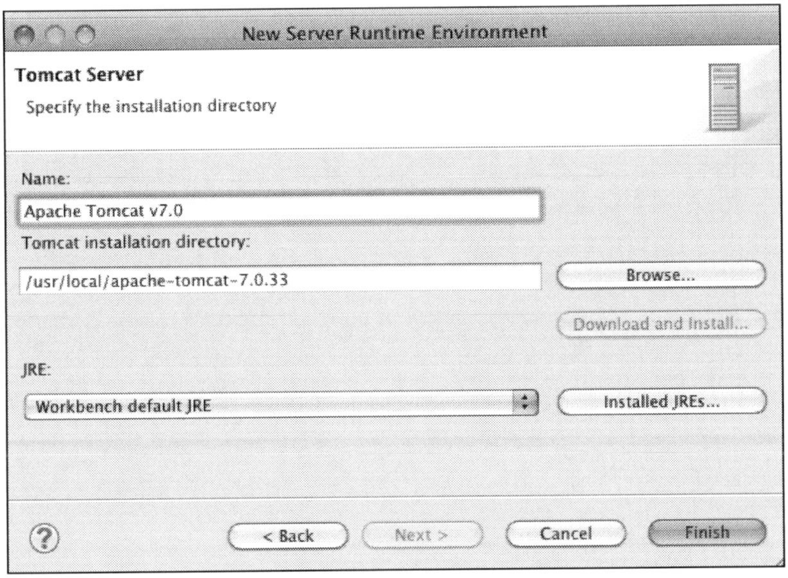

Make sure that the field **Workbench default JRE** (in the bottom of the previous figure) is pointing to the desired JDK version.

The next step is to create a new Tomcat server within Eclipse. To do so, navigate to **Window | Show View | Other...**. Then click on the **Server** option to expand it, as shown in the following screenshot. Select **Servers**, and press **OK**:

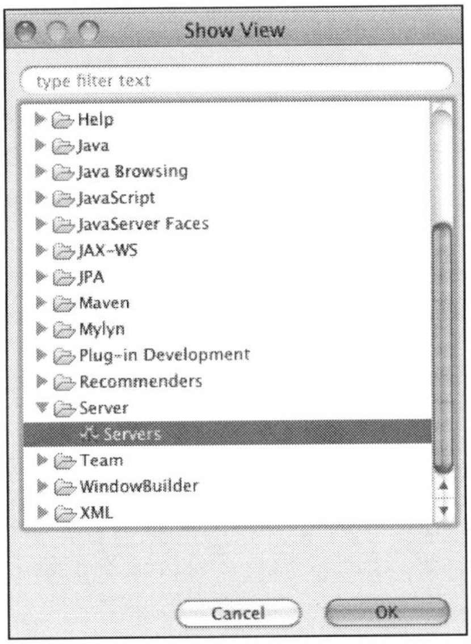

Now, as you can see in the following screenshot, the server view displays the status of Tomcat. Through this view you can start, stop, and restart Tomcat by right-clicking on Tomcat's icon:

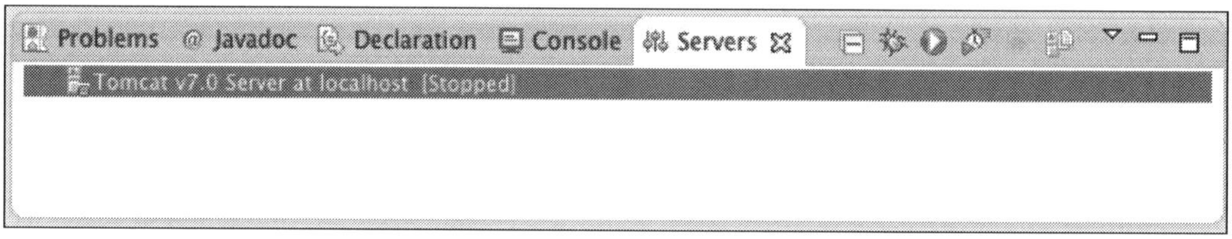

You are now ready to start using Eclipse with Tomcat to develop servlets and JSPs. Nevertheless, before implementing your first servlet, let's take a look at how servlets are executed by servlet containers.

The servlet lifecycle

Before we go on, a few words about the lifecycle of servlets is in order. There are three methods implement the lifecycle of a servlet: `init()`, `service()`, and `destroy()`. Prior to handling client requests, the servlet is instantiated and initialized by the servlet container (Tomcat in our case). Only one instance of each servlet is instantiated by the servlet container. During this initialization phase, the container calls `init()` to give the servlet a chance to run initialization code. Servlets operate in a multithreaded fashion, which means that their methods may be invoked by a number of threads. Upon receiving the first client request, the container starts a new thread or allocates one from the pool. Then, the execution of the servlet is assigned to that thread, which carries on by invoking the servlet's `service()` method.

> It is worth emphasizing that `init()` is executed only once. Also the container is not able to invoke `service()` before `init()` completes.

`service()` takes two parameters: a request object (which is an instance of `javax.servlet.http.HttpServletRequest`) and a response object (which is an instance of `javax.servlet.http.HttpServletResponse`). It is through these two instances that the container accesses the client request and generates output. Within `service()`, the request object is examined to determine what sort of request it is. After determining the request type, `service()` dispatches the request and response objects to the HTTP-related method responsible for implementing the request in question. Two examples of HTTP-related methods are `doGet()` and `doPost()`. These methods correspond to the HTTP `GET` and `POST` operations, respectively.

> In addition to doGet() and doPost(), which are used most of the time, HttpServlet also contains the doPut() and doDelete() methods. These methods correspond to the HTTP PUT and DELETE operations, respectively. But since these two are seldom used, we will not go into detail. It suffices you to know that they exist.

Lastly, the main purpose of destroy() is to give servlets a chance to clean things up. Summing up, most of the time when implementing your own servlets, the three methods that you will have to override are init, doGet, and doPost (and possibly destroy if you need to clean some resources up).

Implementing your very first Java servlet using Eclipse WTP

First, we will explain how to implement servlets using Eclipse WTP, and then we will come around to implementing JSPs later. We will create a quite simple servlet that prints the classic phrase "Hello, World!". You will see that writing and deploying servlets using Eclipse WTP is a breeze. Certain aspects of developing web applications using Eclipse WTP are so easy that you may not need to type almost anything. So, if you are really serious about learning how to develop web applications using servlets and JSP, we recommend you to skim through this chapter just to get a feel of how to use WTP, and then take a break from Eclipse altogether and start arranging all the structure of your web applications by hand. That will help you to better understand how this sort of applications are organized and the number of steps Eclipse has graciously automated for you. After getting the gist of how to assemble, develop, and deploy web applications, you can go back to Eclipse in order to speed up the development process. We direct the interested readers to http://tomcat.apache.org/tomcat-7.0-doc/appdev/index.html for further information on web application development.

To develop servlets and JSPs using Eclipse WTP, first, we need to create a new dynamic web project: go to **File** | **New** | **Other...** | **Web**. In the **New** wizard, expand **Web**, and then select **Dynamic Web Project**. You should reach the wizard shown in the following screenshot.

If you take some time observing the items under the option **Web**, you will see that there are two types of web projects: dynamic and static. Dynamic web projects contain dynamic resources, such as servlets, JSPs, and a bunch of metadata, to name but a few items. Static web projects include only static resources, for instance, cascading stylesheets.

> The Java conventions entail a lot of overhead; you have to comply with a directory structure that specifies the location of class paths, supporting metadata, and deployment descriptors. So, when you want to create a simple web application that has only static content, you might want to use the static web project type.

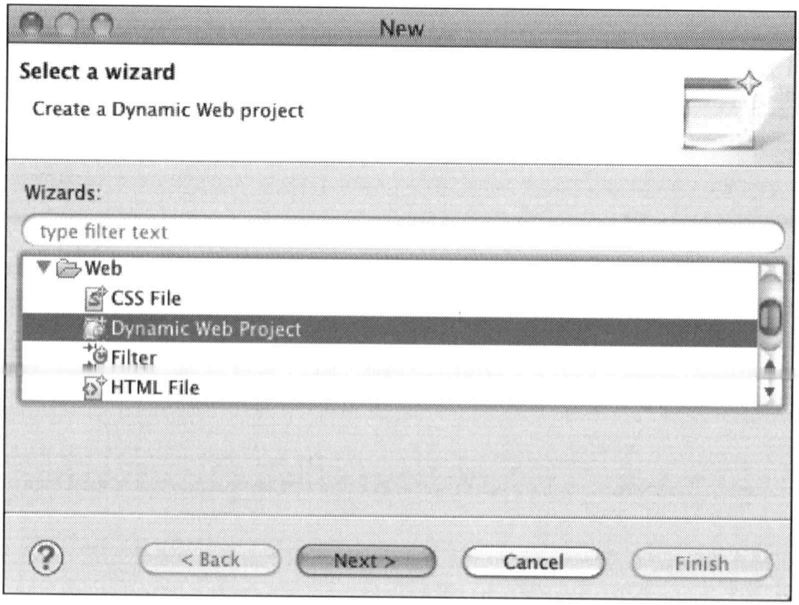

After selecting **Dynamic Web Project**, hit **Next**. Let's name our project `chapter7.examples` (type it in the **Project name** textfield). Some words about the **New Dynamic Web Project** wizard (shown in the following screenshot) are in order. First, as you can see in the following screenshot, it is possible to specify the server you want to use by selecting an option from the **Target runtime** drop-down list. Given that we installed Tomcat, it should be the default option. Second, as for the number under **Dynamic web module version**, it has to do with the servlet API version that will be used throughout the project. As of this writing, 3.0 is the latest version, and the one that we will use in this chapter. So, set this option to **3.0** before clicking on **Next**. Lastly, you can either leave the **Configuration** option set to **<custom>** or **Default Configuration for Apache Tomcat v7.0**.

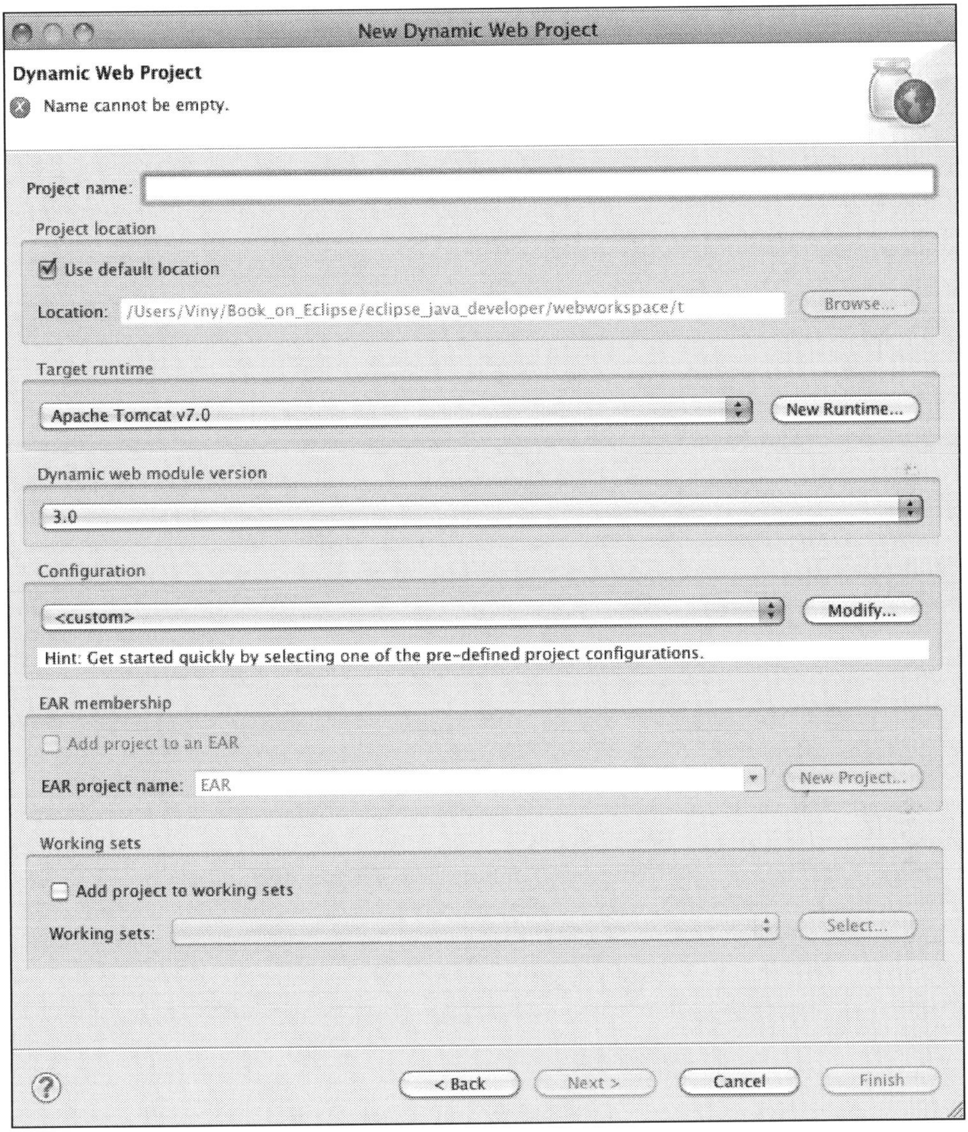

The next window, named **Configure project for building Java application** (this window is not shown), can be used to add folders for your source on the build path. In this window, do not change anything, just accept the default value (`src`), and hit **Next** again.

Finally, the last window asks whether an application deployment description (`web.xml`) should be generated or not. Note that generating this XML file is not the default option. For the time being, we do not need such a file, so leave this option unselected.

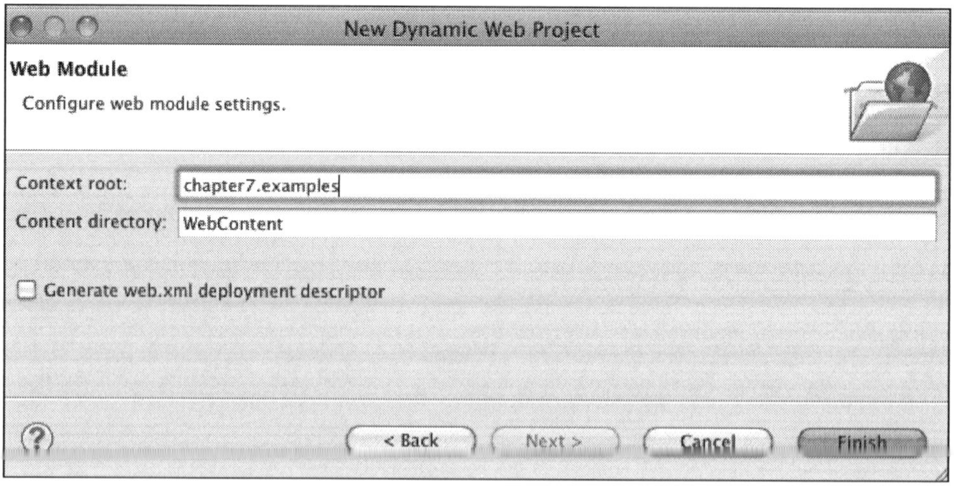

Upon clicking on **Finish**, Eclipse will prompt you to change perspective. Dynamic web projects are associated with the Java EE perspective. To proceed, select **Yes**. You just created a new project that sports the structure of Java web applications. Note that you did not have to create a myriad of folders nor XML files. The next screenshot shows an overview of how dynamic web projects are organized. For the time being, the important parts for us are the folders, `WebContent` and `Java Resources`, which is where your servlet will go. `WebContent` comprises two subfolders: `META-INF` and `WEB-INF`. The latter is of particular importance here. `WEB-INF` contains the application deployment descriptor (`web.xml`, a file that contains configuration information) and subfolders for other sorts of resources, for instance, configuration files. It is important to mention that clients do not have access to the files under this folder, which makes it a safe place for files that are not meant to be publicly available. Inside `WEB-INF` there is a subfolder named `lib`, where you place JAR files. Class files that are not packaged in JAR files are placed into another folder named `classes`. However, it bears mentioning that you should not place any `.class` file into this directory. Instead, these files are placed there automatically when Eclipse's Java compiler translates Java source files that are in the `Java Resources` directory. When the compiler runs, it deletes all files placed directly in this directory.

> Java files in the `classes` folder must be stored in a directory structure that mirrors their package structure. For instance, a class that belongs to a package named `foo.bar` must be stored in the `WEB-INF/classes/foo/bar` folder. When developing inside Eclipse, it takes care of most of these deployment-related issues for you. However, because this structure is common to most Java web applications, knowing where each file should be placed can help you to get your bearings when inspecting someone else's web application.

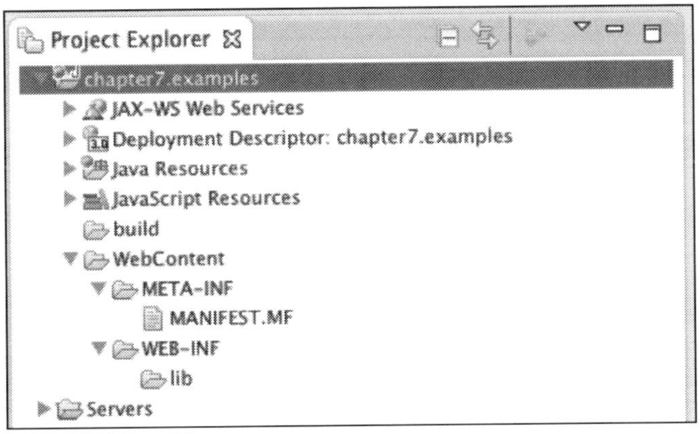

Eclipse has not created any servlet yet. To get Eclipse to create our first servlet, right-click on the folder `WebContent`, and navigate to **New** | **Other...** | **Web** | **Servlet**. Fill the blank fields out, as shown in the following screenshot:

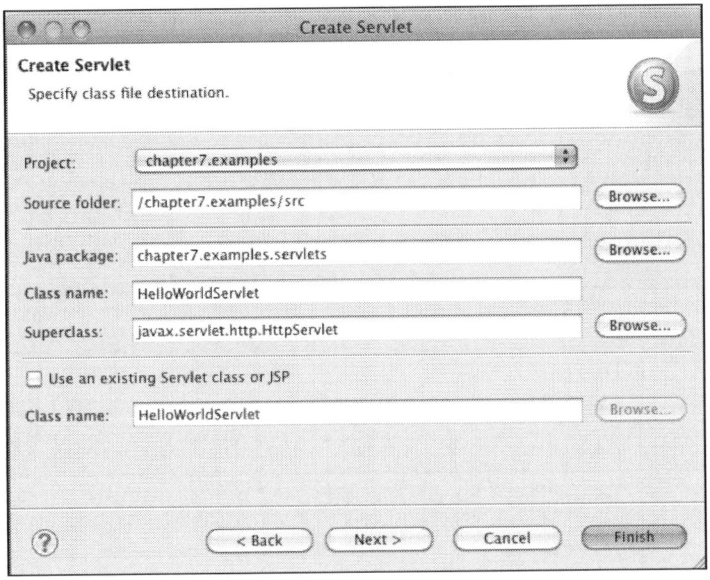

You just named your servlet `HelloWorldServlet` and placed it within a package named `chapter7.examples.servlets`. The generated code is shown as follows. For brevity's sake, most of the generated comments are omitted:

```
package chapter7.examples.servlets;
import java.io.IOException;
import javax.servlet.ServletException;
import javax.servlet.annotation.WebServlet;
import javax.servlet.http.HttpServlet;
import javax.servlet.http.HttpServletRequest;
import javax.servlet.http.HttpServletResponse;
@WebServlet("/HelloWorldServlet")
public class HelloWorldServlet extends HttpServlet {
   private static final long serialVersionUID = 1L;
   public HelloWorldServlet() {
      super();
   }
   protected void doGet(HttpServletRequest request,
HttpServletResponse response) throws ServletException, IOException {
      // TODO Auto-generated method stub
   }
   protected void doPost(HttpServletRequest request,
HttpServletResponse response) throws ServletException, IOException {
      // TODO Auto-generated method stub
   }
}
```

We do not want you to type stuff into your computer mindlessly, so before modifying the generated code to suit our needs, let's go over some of the details of this piece of code. Servlet 3.0 API introduced a lot of new features. One of the new features is the introduction of annotations, which has made the XML file (`web.xml`) that is used to declare servlets—among other things—optional. These annotations reside in the package `javax.servlet.annotation`. Using them you can define metadata regarding a servlet by simply annotating it with `@WebServlet`. You use `@WebServlet` to let Tomcat know that the class in question is a servlet. `@WebServlet` can take a number of parameters, and metadata can be specified for the servlet using such parameters. In this example, `/HelloWorldServlet` represents the URL of the servlet.

> The URL of the servlet can be seen as the name the client knows about. In other words, this is the name that appears in the HTML code accessed by the client—it is a made-up name for clients. It is up to the container to invoke the correct servlet based on the URL provided by the client.

Apart from `@WebServlet`, another characteristic that sort of gives the previous class away as being a servlet is that it extends `javax.servlet.http.HttpServlet`. The auto-generated code also provides stub implementations of the two methods that you will override most of the time when implementing servlets, namely, `doGet` and `doPost`. As you can see, both `doGet` and `doPost` methods take an instance of `HttpServletRequest` and `HttpServletResponse`. These instances provide the input and output streams, respectively, and are used to read and write client data, respectively.

When writing servlets, you will find yourself thinking about the implications of the operation your servlet needs to carry out. In other words, first you need to determine which HTTP-related method conforms to the purpose of the operations you are about to implement. `GET` and `POST` have different semantics. For instance, in HTML parlance, `GET` is a request for retrieving something at a given URL, while `POST` is used to send an arbitrary amount of data to the server. In addition, `GET` operations are idempotent: they can be executed more than once without causing any side effects on the server. `POST` operations, on the other hand, are not idempotent by definition. So, with that in mind, which HTTP-related method do you think we should override? Because `HelloWorldServlet` is only going to show us a message, we can see its implementation as if we were retrieving something: the `Hello, World!` message. Therefore, you guessed right if you thought about overriding `doGet`.

Go ahead and delete `doPost` or comment it out. You will not need it. While you are at it, you can also safely get rid of the auto-generated constructor. Once you have removed them, implement `doGet` as follows:

```
package chapter7.examples.servlets;
import java.io.IOException;
import java.io.PrintWriter;
import javax.servlet.ServletException;
import javax.servlet.annotation.WebServlet;
import javax.servlet.http.HttpServlet;
import javax.servlet.http.HttpServletRequest;
import javax.servlet.http.HttpServletResponse;
@WebServlet("/HelloWorldServlet")
public class HelloWorldServlet extends HttpServlet {
   private static final long serialVersionUID = 1L;
   protected void doGet(HttpServletRequest request,
HttpServletResponse response)   throws ServletException, IOException {
      response.setContentType("text/html");
```

```
        PrintWriter pw = response.getWriter();
        pw.println("<html><head><title>Warm Greeting</title></head>" +
        "<body>" +
            "<h1> Hello, World! </h1>" +
        "</body></html>"
        );
        pw.close();
    }
}
```

Before running your servlet, let's take a look at what you just typed inside `doGet`. First, you should keep in mind that we are overriding `doGet` to handle client requests. Within the context of this example, the `GET` requests to `/HelloWorldServlet` will be answered by an HTML document that prints the message `Hello, World!`. Initially, you need to indicate to the container that the data that is being sent back is an HTML document. To do so, you invoke the method of the `HttpServletResponse` object named `setContentType`. In this example, we set the content type to `text/html`. Information concerning the content type goes in the header of the HTTP response. Such information tells the client what to expect as result from the request. It is through this information that the client browser knows what to do next: render the HTML, launch a PDF viewer, save the returned data as a downloaded file, and so on.

> When you invoke `setContentType`, you are setting the MIME type for the data returned by the servlet. Most of the time, the MIME type used by servlets is `text/html`. However, other MIME types might be used. For instance, a servlet that returns a PDF document should set the MIME type to `application/pdf`.

Through `setContentType` you can also specify the character encoding. Let's say we wanted the servlet container to set the character encoding to UTF-8, we would invoke `setContentType` as follows:

```
response.setContentType("text/html; charset=UTF-8");
```

> From Java 5 onwards, you can also set the character encoding via the servlet response object's `setCharacterEnconding` method.

The next step is to fetch the output stream and write the desired HTML content to it. The servlet response object provides the output stream for sending data back to the client. To get the output stream, you need to invoke `getWriter`, which returns a `PrintWriter` instance. If you have already used `java.io`, then you are already familiar with `PrintWriter`. If you have not used `java.io`, here is a crash course on `PrintWriter`: it is designed to handle character data. When you need to return something else from your servlet, you will use an `OuputStream` instance. You can obtain an `OuputStream` instance by invoking `getOutputStream` on a response object as shown in the following snippet:

```
ServletOutputStream out = response.getOutputStream();
```

`PrintWriter` has a lot of methods; however, most of the time you will be using just `println` or `print` to print text data to the character stream. As shown in the previous example, in order to write HTML to the stream, you simply perform I/O operations. To be more specific, HTML is embedded into Java strings and sent to the output stream.

At this point you should be able to run the recently created servlet. Running our example inside Eclipse is quite straightforward. Select `HelloWorldServlet`, right-click on it, and then select **Run on Server** from **Run As**, as shown in the following screenshot:

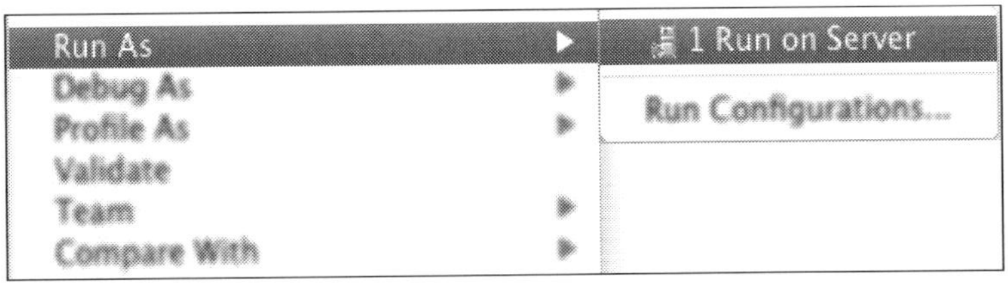

It is worth mentioning that Eclipse also allows you to debug web applications. You can even set breakpoints within JSP files. So if you ever face any problem during development, it is good to know that you can resort to a step-by-step execution in order to uncover bugs.

A window listing the available servers will appear. Because we configured Tomcat previously, it should be included in this list (as shown in the following screenshot). After selecting **Tomcat**, click on **Finish**. You can check the option **Always use this server when running this project**, so that this dialog will not pop up whenever you run your application.

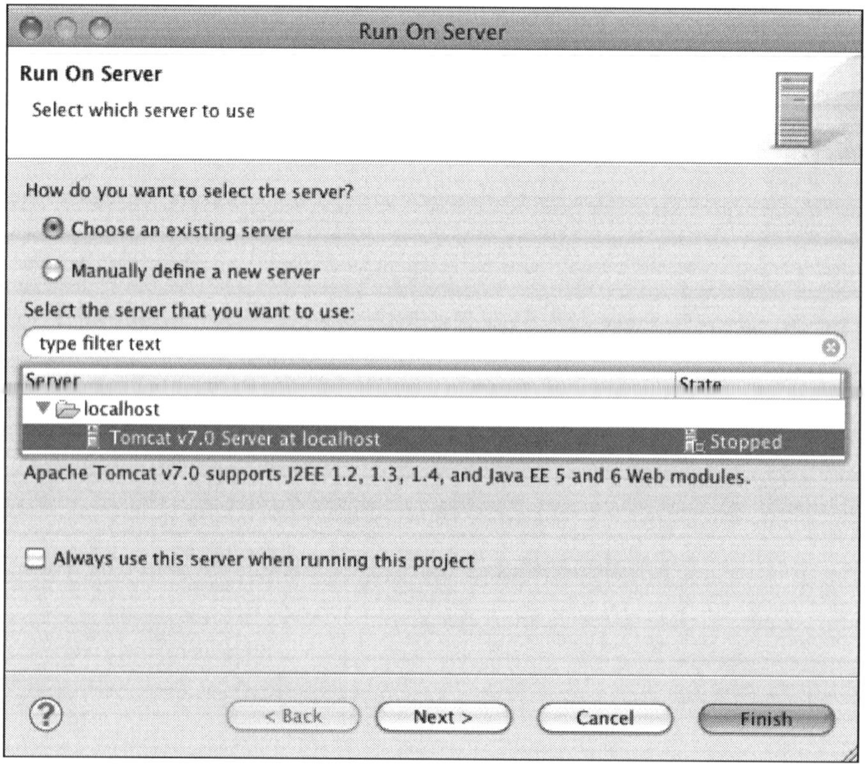

Upon pressing **Finish**, Eclipse will configure the example servlet on Tomcat. You should see the following message being displayed in Eclipse's internal browser. If you got this message, congratulations, you have just created your first servlet using Eclipse WTP:

We can make the message shown by this servlet more interesting. So far, there is nothing dynamic about it. Every time you run the example, you get the same message. In other words, it does not matter how many times you refresh Eclipse's internal browser, you will see the **Hello, World!** message. Let's add the current time to this message. To do so, modify the `doGet` method as shown in the following code. Note that whenever you need to change the HTML data that is sent back to the client you will have to change the HTML data that is embedded in your Java code:

```
protected void doGet(HttpServletRequest request, HttpServletResponse
response) throws ServletException, IOException {
   response.setContentType("text/html");

   java.util.Date currentTime = new java.util.Date();
   PrintWriter pw = response.getWriter();
   pw.println("<html><head><title>Warm Greeting</title></head>" +
      "<body>" +
          "<h1> Hello, World! </h1>" +
      "<ul> <li>current time: " + currentTime + "</li></ul>" +
      "</body></html>"
   );
   pw.close();
}
```

To see the new servlet at work, run it as described before. Upon running the previous code, you should see something similar to the following screenshot. Now, every time you hit **Refresh**, you get a brand new timestamp. This timestamp is obtained during runtime and dynamically added to the response:

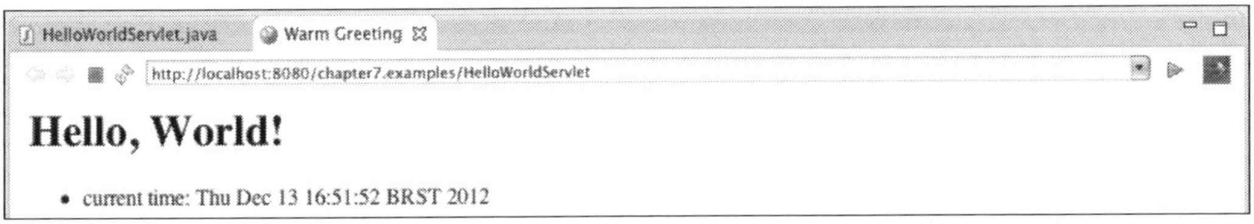

A more elaborate example – yet another calculator

Now that you know the basics, let's move on to a more elaborate example. We are going to develop a simple calculator (by this time, you might have realized that one of the authors is a calculator buff). To keep things simple, our web calculator will be able to deal with only two operands at a time. After typing the value for each operand, the user will be able to select the operation that he or she wants the calculator to compute. The result of the operation will be shown in a new page.

Our example will comprise three classes, of which two are servlets and one is a plain Java class. The plain class implements our calculator itself, while the servlets are responsible for the presentation and the interplay between the presentation and the calculator class. Let's start small by implementing the first servlet, the one in which the user will type the value of the two operands. It bears mentioning that we could have implemented the same functionality using plain HTML, but we like to make things more interesting. Moreover, implementing this functionality as a servlet will give you another opportunity to put your recently acquired knowledge into practice. Okay, let's get down to business.

We are going to develop our calculator in a new project. So, first, create a new dynamic web project, and name it `webcalculator` (if you do not remember how to create a dynamic web project, go back to the beginning of the section in which we explain how to create servlets in Eclipse). Then, create a servlet named `OperationSelect` and place it in the package `com.example.calculator.web`. The next screenshot shows what you should type in the **Create Servlet** window. Remember that you do not need to create the package beforehand. Upon creating your servlet, Eclipse makes sure to create its associated package structure. Because our project comprises both servlets and plain Java classes, we will use a widely used convention to organize these elements. Servlets will be placed in the package `com.example.calculator.web`, while our calculator class will be placed in a package named `com.example.calculator.model`.

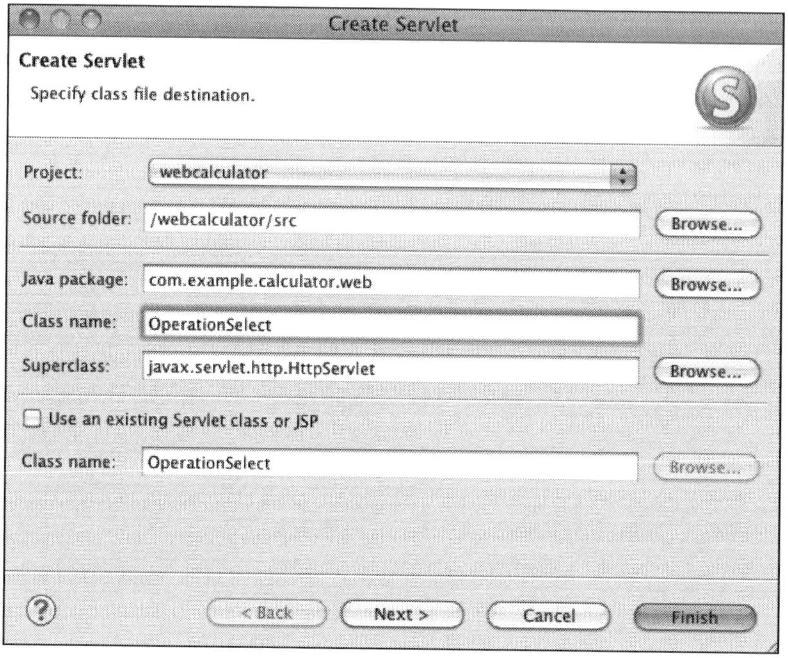

Click on **Next** twice so that you can get to the following window. Then unselect the **doPost** option; we will not override this method for this servlet. While you are at it, also unselect the **Constructors from superclass** option; we do not need a constructor either. Lastly, hit **Finish**:

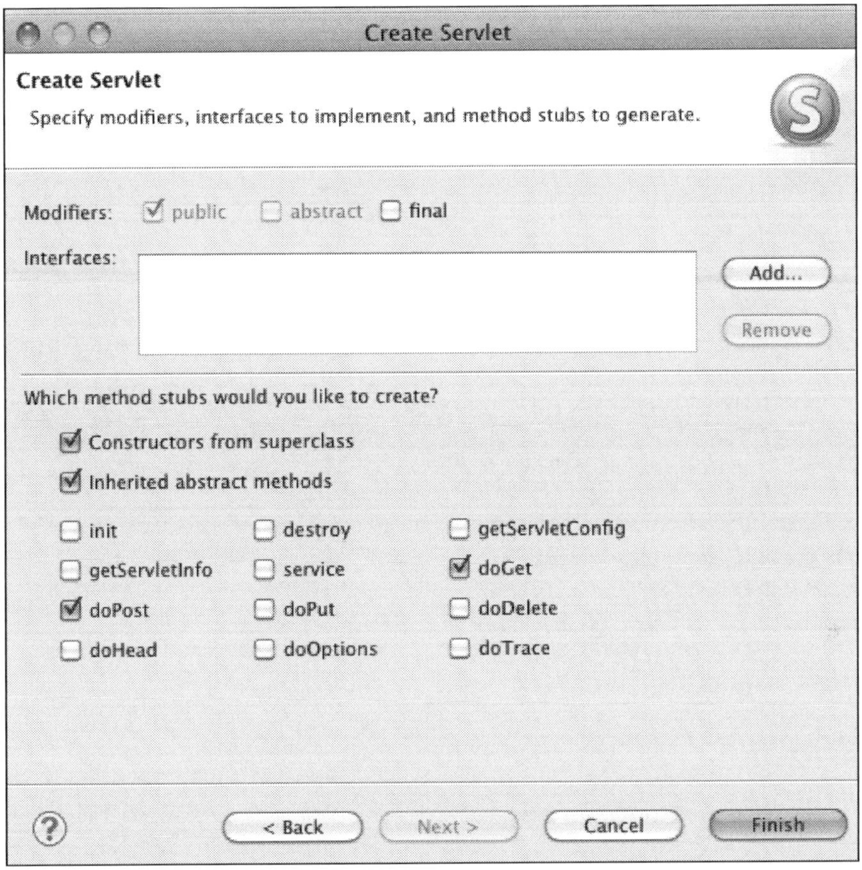

Eclipse will generate a servlet containing an empty `doGet` method. Although the generated code does do much, we do not need to type everything from scratch now. Let's start changing our generated servlet by modifying the `doGet` method, as shown in the following code. Note that we are performing the same operations shown in the previous example. First, we set the content type to `text/html`. Second, we fetched the output stream in order to write our HTML content to the output.

As you can see, the more HTML you have to add to the output, the more unwieldy your code becomes. For the sake of simplicity, we did not use complex HTML constructs inside our example servlet. The first two elements are textfields, and the last one is a drop-down list from which the user can select the operation to be performed using the two operators. We also add a `Submit` button that will forward the operands and operation information to another servlet that, in turn, will trigger the execution of our calculator class:

```
@Override
protected void doGet(HttpServletRequest request, HttpServletResponse
response) throws ServletException, IOException {
    response.setContentType("text/html");
    PrintWriter pw = response.getWriter();
    pw.write("<html><body>");
    pw.write("<h1 allign=\"center\">Web Calculator v1.0</h1>");
    pw.write("<form method=\"POST\" action=\"PerformCalculation.
do\">");
    pw.write("<tt>1st operand:</tt> <input type=\"text\"" +
"name=\"firstOperand\"><br>");
    pw.write("<tt>2nd operand:</tt> <input type=\"text\"" +
"name=\"secondOperand\"><br>");
    pw.write("Operation to be performed: ");
    pw.write("<select name=\"operation\" size=\"1\">");
    pw.write("<option value=\"+\"> + </option>");
    pw.write("<option value=\"-\"> - </option>");
    pw.write("<option value=\"*\"> * </option>");
    pw.write("<option value=\"/\"> / </option>");
    pw.write("</select></br></br>");
    pw.write("<input type=\"submit\" value=\"Submit\">");
    pw.write("</form>");
    pw.write("</body></html>");
    pw.close();
}
```

Deploying the calculator example

After typing the previously shown method, we are ready to take the first part of our calculator example for a spin. In case you do not remember how to run web applications inside Eclipse, here is a quick reminder: select the desired program (in this case, `OperationSelect`), right-click on it, and then navigate to **Run on Server** in **Run As**. Select **Tomcat** from the list of available servers, and click on **Finish**. The next screenshot shows what you will see after executing this servlet:

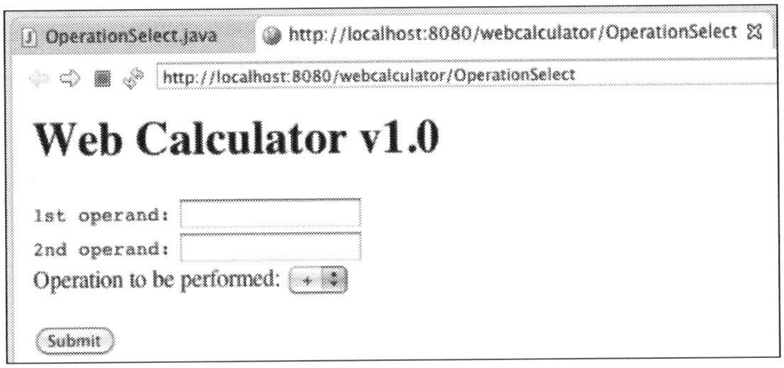

We have not implemented the servlet responsible for forwarding the values and operation to the class that implements the calculator. Therefore, if you type two numbers in the **1st operand** and **2nd operand** fields, select an operation, and press **Submit**, you will get the page shown in the next screenshot. This indicates that the resource we are looking for was not found by the servlet container. So, let's take a step back and see why the servlet container is trying to forward the data to `webcalculator/PerformCalculation.do`.

The answer to this question boils down to the fact that our form is submitting information to something named `PerformCalculation.do` (take a look at the previous chunk of code). Further, given that there is no resource (servlet or HTML page, for instance) with such a name, the container shows an error message. In this example, `PerformCalculation.do` is a logical name. That is to say, it is no actual filename. We do not want the client to have access to the names of our servlet classes. So, we use made-up names that are mapped to actual servlets instead. That explains why we got an error page: we have not informed the container yet about which servlet answers by the logical name `PerformCalculation.do`.

> The `.do` extension (which implies *do something*) is just a convention.

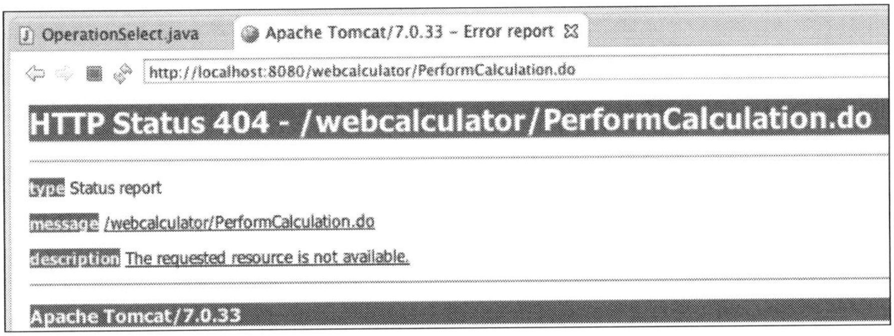

Implementing the second part of our example entails mapping a servlet to the aforementioned made-up name. Again, use Eclipse to generate the servlet for you. Name it `ArithmeticExpert`, and place it into the same package of the previous servlet (`com.example.calculator.web`). When creating `ArithmeticExpert`, and after going through the first window where you type the servlet's name and package, click on **Next**, and then click on **Edit...** as shown in the following screenshot. Here you can edit the servlet URL mapping. Given that we want to inform the container that the new servlet can be mapped to `PerformCalculation.do`, edit this option as shown in the following code snippet (make sure you do not start with a / character) and press **Next**:

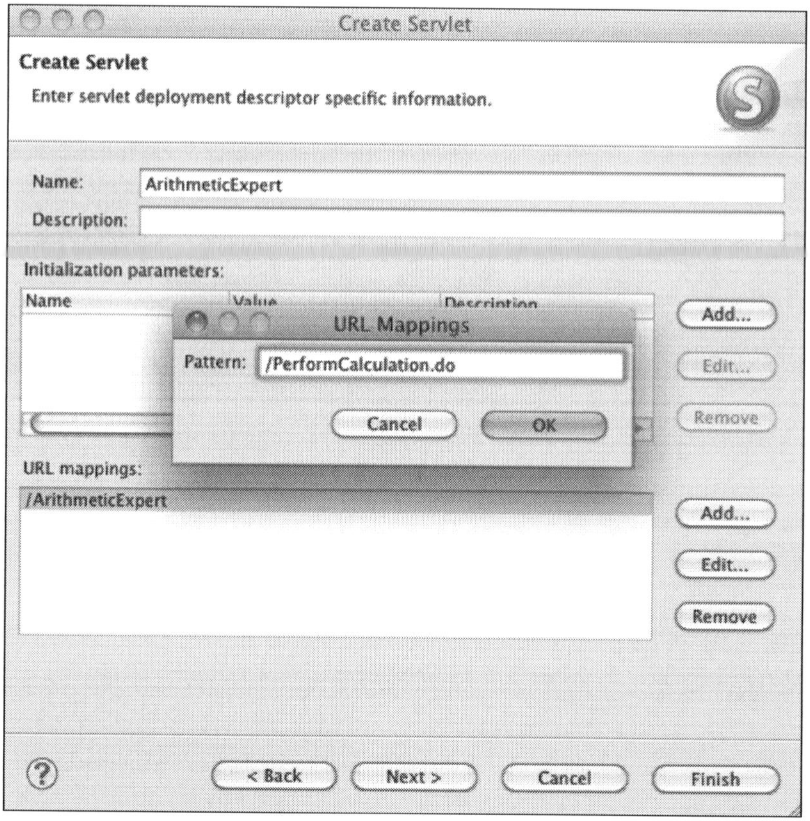

Rather than overriding `doGet`, this time we will override `doPost`. Because the form in `OperationSelect` generates a `POST` operation, we have to override `doPost`. Otherwise, pressing the **Submit** button will not invoke our method. Therefore, unselect the **doGet** and **Constructors from superclass** options. Lastly, hit **Finish**. Change the generated code, as shown in the following snippet:

```
@WebServlet("/PerformCalculation.do")
public class ArithmeticExpert extends HttpServlet {
  private static final long serialVersionUID = 1L;
  @Override
  protected void doPost(HttpServletRequest request,
HttpServletResponse response)
      throws ServletException, IOException {
    response.setContentType("text/html");
    PrintWriter pw = response.getWriter();
    pw.write("<html><body>");
    pw.write("The result of " +
   request.getParameter("firstOperand") + " " +
   request.getParameter("operation") + " " +
   request.getParameter("secondOperand") +
   " is surprise, surprise");
    pw.write("</body></html>");
    pw.close();
  }
}
```

We are implementing our calculator example in stages. In this first version, our goal is to make sure that the communication link between the first and second servlets is working properly. Therefore, our code does not invoke the calculator class to perform the selected operation. Instead, it will just print out a message indicating that the communication between both servlets is working and that we are able to share information between them. There are two important points to note in the preceding code. First, see how the mapping between the servlet and its made-up name, which appears in the output of `OperationSelect`, is implemented using the `@WebServlet` annotation. Note that the string enclosed in parentheses is the pattern we typed during the creation of the underlying servlet (check the previous screenshot). Second, note how we access the parameters from the request. We use the method `getParameter` from `HttpServletRequest` (actually, this method is defined in the interface `ServletRequest`, and implemented by subclasses including `HttpServletRequest`). Using `getParameter`, each parameter is retrieved by its name.

> Parameter names are case sensitive.

Upon running the previous servlet and typing `2` in the **1st operand** textfield, `1` in the **2nd operand** textfield, selecting subtraction as the operation to be performed, and pressing **Submit**, you should see something like the next screenshot. As you can see in the following screenshot, now the application can access data entered by the user in one of the servlets to perform the calculations in the other servlet:

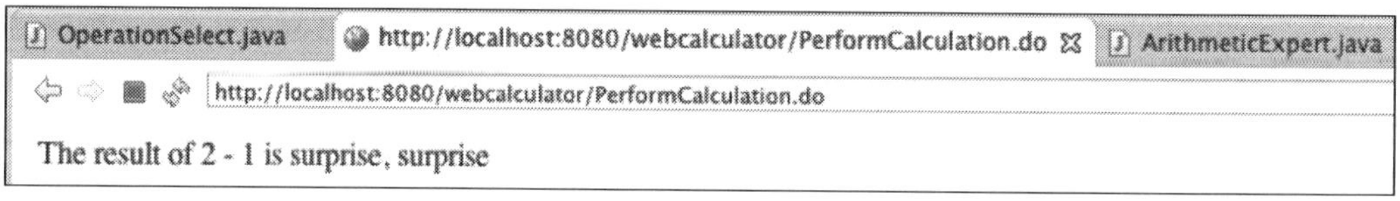

To finish the calculator example, we need to implement a calculator class that is able to perform the arithmetic operations shown in the first servlet. That should be straightforward for you now. Before checking our solution shown in the following snippet, try to write this class by yourself. Just make sure to name it `Calculator`, and place it into a package named `com.example.calculator.model`. The following is how we implemented `Calculator`:

```
package com.example.calculator.model;
public class Calculator {
  private double op1, op2;
  private String operation;
  public Calculator(String op1, String op2, String operation) {
    try {
      this.op1 = Double.parseDouble(op1);
      this.op2 = Double.parseDouble(op2);
      this.operation = operation;
    }
    catch (NumberFormatException e) {
      //reset the value of both operands
      this.op1 = 0.0;
      this.op2 = 0.0;
    }
  }
  public double performCalculation() {
    if (operation.equals("+")) {
      return op1 + op2;
    } else if (operation.equals("-")) {
      return op1 - op2;
```

```
    } else if (operation.equals("/")) {
      return op1 / op2;
    } else if (operation.equals("*")) {
      return op1 * op2;
    }
    return 0;
  }
```

After implementing `Calculator`, you just need to integrate it to the `ArithmeticExpert` servlet, as shown in the following snippet. Note that our implementation does not deal with edge cases as division by zero. Upon executing the example again, you will see something similar to what is shown in the following screenshot:

```
@Override
protected void doPost(HttpServletRequest request, HttpServletResponse 
response) throws
   ServletException, IOException {
  response.setContentType("text/html");
  PrintWriter pw = response.getWriter();
  pw.write("<html><body>");
  String operandOne = request.getParameter("firstOperand");
  String operandTwo = request.getParameter("secondOperand");
  String operation = request.getParameter("operation");
  pw.write("The result of " +
   operandOne + " " +
   operation + " " +
   operandTwo +
   " is " +
   new Calculator(operandOne, operandTwo,
     operation).performCalculation());
pw.write("</body></html>");
pw.close();
}
```

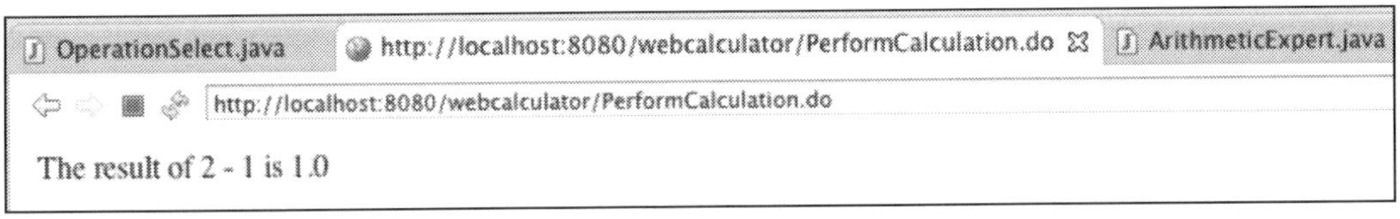

Implementing your very first JSP using Eclipse WTP

The previous sections got us started with Tomcat, servlets, and Eclipse WTP. As you saw, despite the benefits provided by Eclipse WTP (it generated all the structure of the web project for us and also took care of generating a significant chunk of code when we implemented the previous example servlet), embedding HTML into your Java code feels a little awkward. As we mentioned, JSPs were introduced to make the creation of dynamic content more straightforward and appealing. JSPs let you mix HTML and Java code in a better way than servlets. Rather than embedding HTML into Java code, JSP makes it possible to embed dynamic content into HTML. Recall from the previous sections that a JSP is an HTML page containing some special elements. The elements that are rendered into dynamic content resemble HTML elements. However, they are componentized Java programs. Besides these HTML-like elements, a JSP page can also include Java code (*scriptlets*).

It bears reminding you of the fact that although JSPs look like standard HTML pages with additional elements, they are compiled down to servlets. As we mentioned, at runtime, Tomcat compiles JSP pages down to servlets, compiles the servlets to bytecodes, and executes them eventually.

In this section we will demonstrate how to use Eclipse WTP to create JSPs. We will implement a JSP that shows the same message of the servlet example.

You do not need to create a new web application project; we will use the previous one. To get Eclipse to create a new JSP file for you, do the following: select the folder `WebContent` (Tomcat will look for JSPs in this folder), and right-click on **JSP File** in **New**. Name it `HelloWorldJSP.jsp` (as shown in the following screenshot), and then hit **Next**. Actually, you do not need to type the extension (`.jsp`), Eclipse will include it automatically:

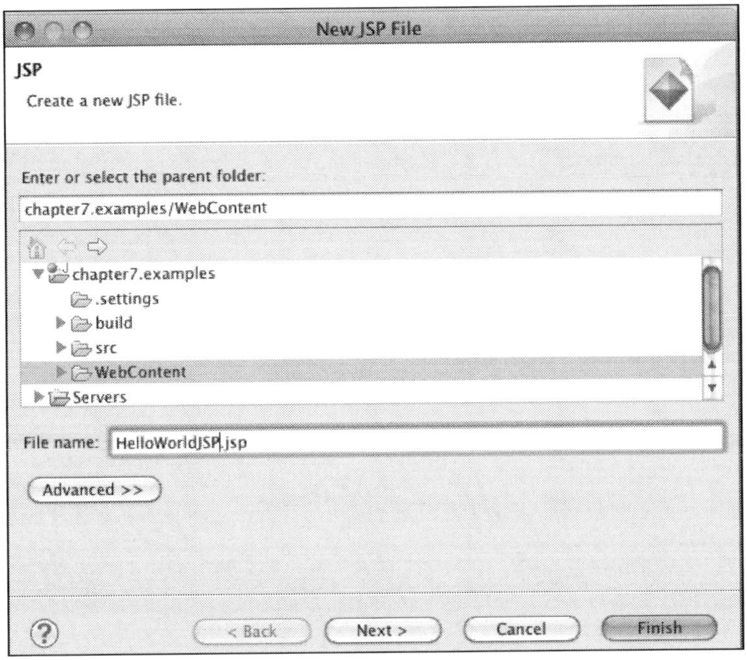

Afterwards, select the option named **New JSP File (html)**, whose description is **JSP with html markup** (as illustrated in the following screenshot), and click on **Finish**:

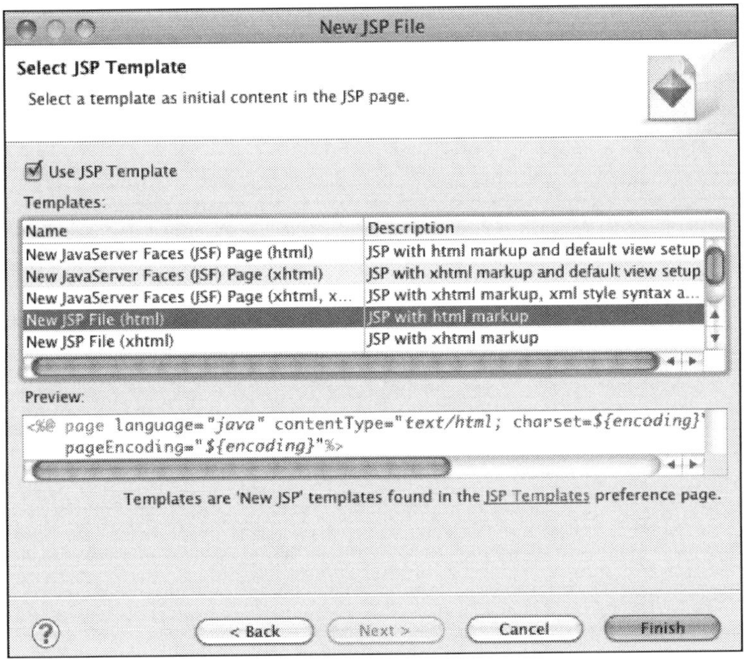

Eclipse will generate the following code. The first line is a JSP directive element. Directive elements specify various attributes related to the page itself. Thus, it does not necessarily change the response sent to the browser, but it informs the container how it should deal with the page. The first line is a page directive that, among other things, specifies the content type for the page in question: `html/text`. JSP directive elements start with `<%@` and end with `%>`, and between these two delimiters a list of name/value pairs represents the so-called directive attributes.

Apart from the page directive, the auto-generated code contains mostly HTML code. In JSP parlance, everything that is not a JSP element is named template text. Template text is static, which means that it is sent to the browser without undergoing any transformation.

As for the template text in the auto-generated code, there is a title placeholder and the body part is empty. If you know HTML and Java, writing JSP pages should be a breeze for you. If you are not that familiar with HTML, creating JSP pages may be a little tricky at first.

```
<%@ page language="java" contentType="text/html; charset=ISO-8859-1"
  pageEncoding="ISO-8859-1"%>
<!DOCTYPE html PUBLIC "-//W3C//DTD HTML 4.01 Transitional//EN"
"http://www.w3.org/TR/html4/loose.dtd">
<html>
<head>
<meta http-equiv="Content-Type" content="text/html;
charset=ISO-8859-1">
<title>Insert title here</title>
</head>
<body>
</body>
</html>
```

Let's start by replacing the title placeholder by the title we used in the previous example, namely, "Warm Greeting":

```
<title>Warm Greeting</title>
```

Next, let's add the `Hello, World!` static message as well as the dynamic part of the message: the timestamp. Adding the former is straightforward, just copy part of the HTML code shown in the example servlet. To include the latter, we are going to embed a Java expression in our page by putting them between `<%=` and `%>` as shown in the following code snippet:

```
<%@ page language="java" contentType="text/html; charset=ISO-8859-1"
    pageEncoding="ISO-8859-1"%>
<!DOCTYPE html PUBLIC "-//W3C//DTD HTML 4.01 Transitional//EN"
"http://www.w3.org/TR/html4/loose.dtd">
<html>
<head>
<meta http-equiv="Content-Type" content="text/html;
charset=ISO-8859-1">
<title>Warm Greeting</title>
</head>
<body>
<h1> Hello, World! </h1> <%-- title as template text --%>
<%-- timestamp as embedded expression --%>
<ul> <li>current time: <%= new java.util.Date() %> </li></ul>
</body>
</html>
```

If you run the preceding example, you will get exactly the same message that you got when you executed the example servlet. Now, let's go over the most important parts of the chunk of the preceding code. First, as you might have noticed, everything placed within `<%--` and `--%>` is ignored by the container. You can use this to describe what is going on in your code or to comment out parts of the code. Given that JSP comments are also JSP elements, they never make it to the browser. Second, instead of concatenating a string and using plain I/O to write to the response, the preceding code uses a scripting element through which you can add Java code to JSPs. When you want to simply add Java code without writing the result to the response, you use `<%` and `%>`. When the result needs to be added to the response, you use `<%=` and `%>`. Thus, we used the latter to write the timestamp to the response. The next example shows yet another way to achieve the same thing, but this time using more JSP elements and scriptlets. Note that the page directive is also used to import java classes (and packages), as shown in the following example. In addition, when importing more than one class, the list of fully-qualified names must be separated by commas. After adding import instructions to the page directive, you do not need to use the fully qualified class name inside scriptlets.

> You can also print stuff within a `<%` and `%>` pair. To do so, you have to use the out object. This object is always implicitly there. For example, you could print the `Hello, World!` message as: `<% out.println("Hello, World!") %>`. Nevertheless, it bears mentioning that using `out.println()` inside a scriptlet is a bad practice. Whenever possible, use an expression (`<%=` and `%>`) instead.

```
<%@ page language="java" contentType="text/html; charset=ISO-8859-1"
  pageEncoding="ISO-8859-1" import="java.util.Date" %>
<!DOCTYPE html PUBLIC "-//W3C//DTD HTML 4.01 Transitional//EN"
"http://www.w3.org/TR/html4/loose.dtd">
<html>
<head>
<meta http-equiv="Content-Type" content="text/html;
charset=ISO-8859-1">
<title>Warm Greeting</title>
</head>
<body>
<h1> <% out.println("Hello, World!"); %> </h1>
<% Date date = new Date(); %>
<ul> <li>current time: <%= date %> </li></ul>
</body>
</html>
```

> Never end expressions with a semicolon. `<%= date %>` is correct, `<%= date; %>` is not.

Generating WAR files

Java web applications are distributed in **WAR (Web application Archive)** files. A WAR file is a compressed file that keeps the structure of web applications (apart from the web application context directory, which is the application's top directory), and includes one extra folder named `META-INF`. These compressed files have a `.war` extension. The `META-INF` folder contains metadata about the files within a WAR file. An important file kept in this folder is the `MANIFEST.MF` file. Among other things, such a special file has information about the files packaged in a WAR file.

When WAR files are deployed, they are unpacked by the container, which creates the structure shown in the following figure. The name of the top directory depends on the container in question. Tomcat, for instance, names the top directory after the WAR file. So, for instance, the following structure would be created by Tomcat upon unpacking a `deployedapp.war` file. As you can see, the internal structure of a WAR file is pretty much the same as the structure of your Eclipse dynamic web project:

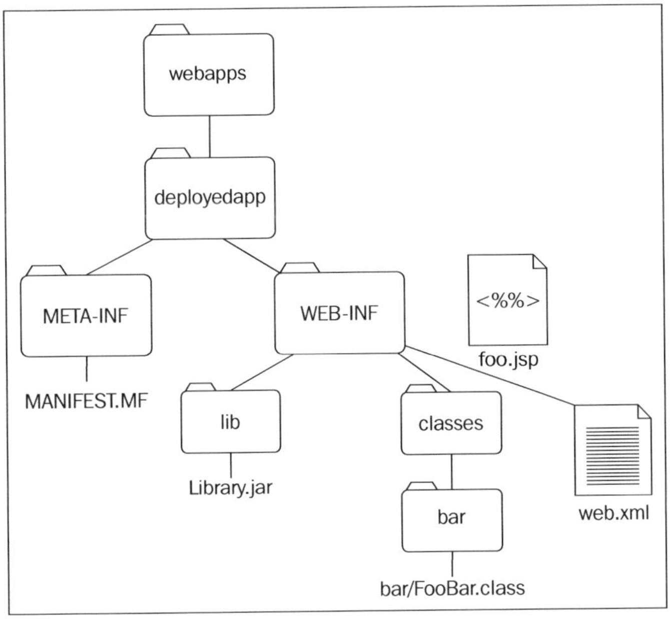

To create a WAR file from your project, right-click on the project and select **Export**. Select the WAR file option as shown in the next screenshot, specify the target directory through the button **Browse...** and hit **Finish**. Now you can deploy your WAR file to Tomcat by placing it under the `webapp` directory and restarting the container.

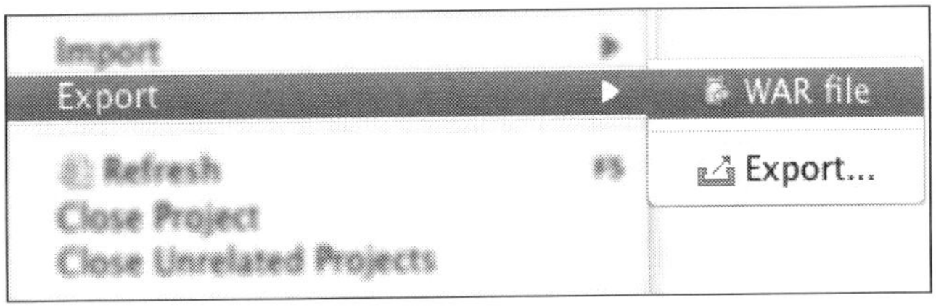

Summary

In this chapter, we presented an overview of the main technologies used to develop Java web applications. We also highlighted some of the advantages of these technologies. This chapter also introduced an important element for these technologies: the container. In particular, the container used throughout this chapter was Tomcat. You saw how to set up Tomcat, as well as how to perform some basic operations as starting and stopping it; both from the command line and from within Eclipse.

You learned that to get Eclipse to support the development of web applications you need to install a set of plugins collectively named Eclipse WTP. In fact, as you saw, you do not need to install the whole set, this chapter explained what plugins you must install.

After learning about the lifecycle of servlets and the role that the container plays in this context, using the installed set of plugins, we went over the creation of a basic servlet and a more elaborate example. While we were at it, we explained the differences between `doGet` and `doPost`. So, now you know when to override each one of them. One of the advantages of using the new servlet API is the use of annotations. Annotating servlets with `@WebServlet` obviates the need of editing the `web.xml` file. Further, you learned that servlets can be tested from within Eclipse by using its internal browser: you just need to properly configure a servlet container. You also learned how to create JSPs using Eclipse WTP. Finally, this chapter showed how to create WAR files from Eclipse web projects. From next chapter on, we will delve into more advanced topics. To be more specific, next chapter addresses the development of plugins using Eclipse.

8
Eclipse Development

In the previous chapters of this book, we have seen how to use Eclipse in various ways, such as developing code for Java, managing projects with Git, and developing web applications. In this chapter, we'll switch the focus and learn how we can use Eclipse to develop code for Eclipse. Since Eclipse is developed in Java, good knowledge of the language is a requisite for this chapter.

One of the Eclipse's features that allowed it to become one of the most popular IDEs around is its pluggable architecture. It contains a runtime platform, which provides the necessary infrastructure for the IDE to run, such as the primary UI elements, a registry of the loaded plugins, a logging infrastructure, among others, and basically everything else is plugins. You can check that by taking a look at the Eclipse's folder in your machine. Inside the plugins folder there's a number of JAR files and folders, each one relating to a plugin. Plugins can be packed either in a simple folder or in a JAR file (which is simply a renamed zip file). Each of these plugins contribute to some aspect of Eclipse. The `org.eclipse.jdt.*` ones, for example, contribute with the Java development tools, such as the Java debugging, the Java editor, and the Java project type.

In Eclipse's `home` folder, you will also find a folder called `features`. A feature is basically a set of plugins. You can use features to group plugins that only make sense when they are bundled together. Take a look at the `feature.xml` file inside the `features/org.eclipse.jdt_<version>` folder. It contains a list of all the plugins this feature contains.

Features and plugins can have dependencies between them. The `org.eclipse.pde` feature, for example, which provides the tools for plugin development, depends on the `org.eclipse.jdt` feature, since all Eclipse plugins are written in Java language.

You can see a list of all features and plugins installed in your Eclipse instance by navigating to **Help | About Eclipse SDK**, and clicking on **Installation Details** in the window that appears. The **Installed Software** tab lists all the feature groups installed, and the **Installation History** tab displays when each additional plugin or feature was installed. The **Features** and **Plug-ins** tabs contain information about the plugins and features provider, version, and ID. You can also check under which license the plugins have been released by clicking on the **Legal Info** button, or the **License** button in the **Features** tab.

By the end of this chapter, you will be able to do the following:

- Create a new plugin project
- Run and debug a plugin project
- Declare and implement extension points
- Create custom entries for Eclipse toolbars and menus
- Use the Plugin Spy feature
- Create new Eclipse views
- Export your plugins

Creating your first plugin

Since most of the features that Eclipse contains are provided by plugins, a good way of starting to study Eclipse development is to learn how to create your own plugin. The best way of doing so is by using the Eclipse's new plugin project wizard. Select **File | New | Other** to see a list of wizards. Select the **Plug-in Project** option and click on **Next**. The first window of the wizard contains basic project configuration. The default options are just fine for us. Pick a nice name for your project, such as "HelloWorldPlugin", and click on **Next**.

The next page contains more plugin properties. Again, default options will do, so just click on **Next**. The following page allows you to build your plugin from a template instead of starting it from scratch. Let's select the **Hello, World** template, which creates the plugin's basic file structure and adds a button to the toolbar that displays **Hello, Eclipse World** when pressed. Click on **Finish**. You will be asked if you want to open the Plugin Development perspective. This perspective is very similar to the Java one, with some views specific to plugin development, such as the plug-ins' view.

Before we try running this project, let's take a look at the file structure that was generated. The following screenshot displays the Package Explorer view of the **HelloWorldPlugin** project:

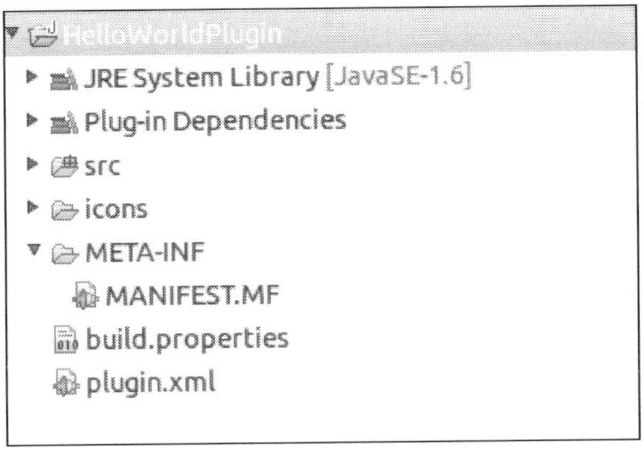

The `MANIFEST.MF` file contains basic information about the plugin. It contains, among others, the plugin's name and version, and information regarding what the plugin provides and on what it depends. This file's format and what it provides are determined by the OSGI specification.

The `plugin.xml` file is another important file in the plugin's basic file structure. It contains information about which extension points the plugin provides and which ones it implements. Extension points are, in general lines, the mechanism through which plugins "talk" to each other. We'll talk more about extension points in this chapter.

The `build.properties` file contains information about which folders contain source files, and which ones should be included in binary and source builds.

By double-clicking any of these three files, Eclipse will open the Plugin Manifest Editor instead of a plain text editor with the file's contents. This editor can be used to manage every aspect of the plugin: dependencies, extension points, exported packages, among others. The editor will, under the hood, edit the `MANIFEST.MF`, `build.properties`, and `plugin.xml` files to add the information that you're providing through its forms. Some Eclipse development guides prefer to present plugin management by showing how to edit these files directly. This chapter, however, will always use the editor's forms, since they are more friendly for the newcomer Eclipse developer.

The following screenshot shows how the Plugin Manifest Editor looks:

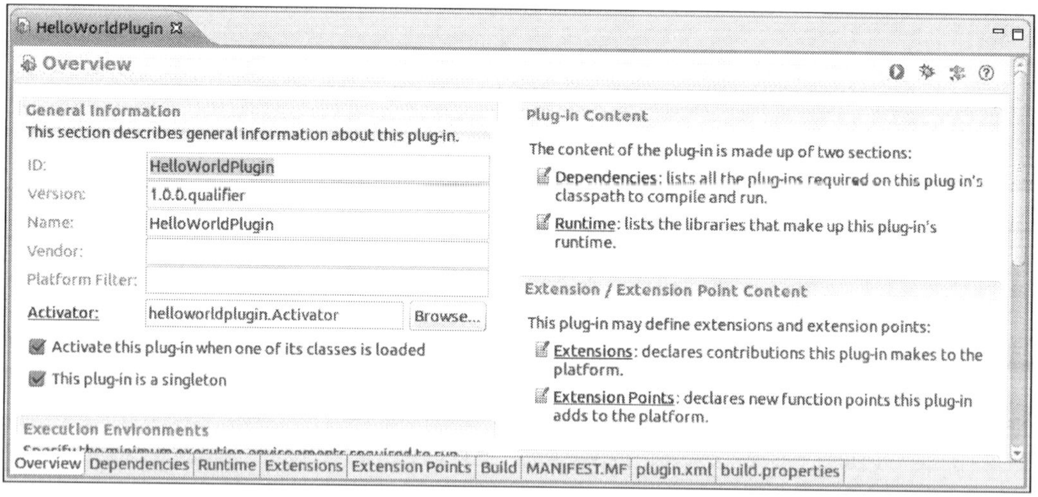

As you can see, there are horizontal tabs in the lower part of the editor. Each one will allow you to control a different aspect of your plugin. The first tab contains basic information about the plugin, as well as links to the other tabs and to plugin development wizards. Now take a look at the **Dependencies** tab, for example. We can see that our plugin depends on **org.eclipse.ui** and **org.eclipse.runtime**, since they are in the list of required plugins. We will describe each of these tabs as we utilize them. The last three tabs allow you to see the actual content of the three plugin's configuration files.

The `src` folder contains two source files: `Activator.java` and `SampleAction.java`, which is the class that effectively displays the message when the button is pressed.

The `activator` class manages the lifecycle of the plugin. It contains code that is executed when the plugin is started or stopped in the Eclipse platform, among other lifecycle-related actions. The `Activator.java` class generated by the wizard contains everything that we need.

Running and debugging a plugin project

Running or debugging a plugin project is not different than how we do it with a Java project. Simply click on the green play button in the toolbar to give a go to our **Hello World** plugin. You will see that a new Eclipse instance will start. By default, this Eclipse instance contains all of the plugins installed in your development environment, plus every plugin project that's open in your workspace. This test instance is called Runtime Workbench.

Look for an Eclipse symbol in the toolbar of your runtime workbench. Click on it, and you'll be greeted with the **Hello, Eclipse World** message.

Debugging works just the same. Stop the current runtime workbench by clicking on the red square button in the console or by closing the Eclipse window. Place a breakpoint in the `SampleAction` class (on line 59), and hit the **Debug** button in the toolbar (the one with a green bug). Now, click on our **Hello World** button in the runtime workbench. The execution will be halted in the selected line. Everything that was discussed in *Chapter 3, Unit Testing with JUnit and Debugging* about debugging is also valid here.

Running and debugging configurations

Eclipse provides further configurations for running and debugging plugin applications. In the top menu, select **Run | Run Configurations** to take a closer look at them.

In the **Main** tab, you will be able to select the workspace location for the runtime workbench. That's where the projects you create while testing and debugging your application will be stored. You can choose to clear the workspace before starting the runtime workbench by clicking on the **Clear** option beneath the workspace location. The **Program to Run** section in the **Main** tab allows you to choose a different product or application to run, and **Java Runtime Environment** allows you to use a different JRE for the runtime environment. This can be useful if you are testing your plugin's compatibility with different JRE's.

The **Arguments** tab allows you to choose arguments both for Eclipse and for JVM to be used when launching the runtime workbench.

The plugins tab allows you to select the plugins that will be loaded in the runtime workbench, both from your Workspace and from your development environment. This can be useful to test your workspace projects individually without having to close them, or to remove the unrelated plugins from the development environment to save your machine's resources. The following screenshot displays the Plugins tab of the **Run Configurations** menu:

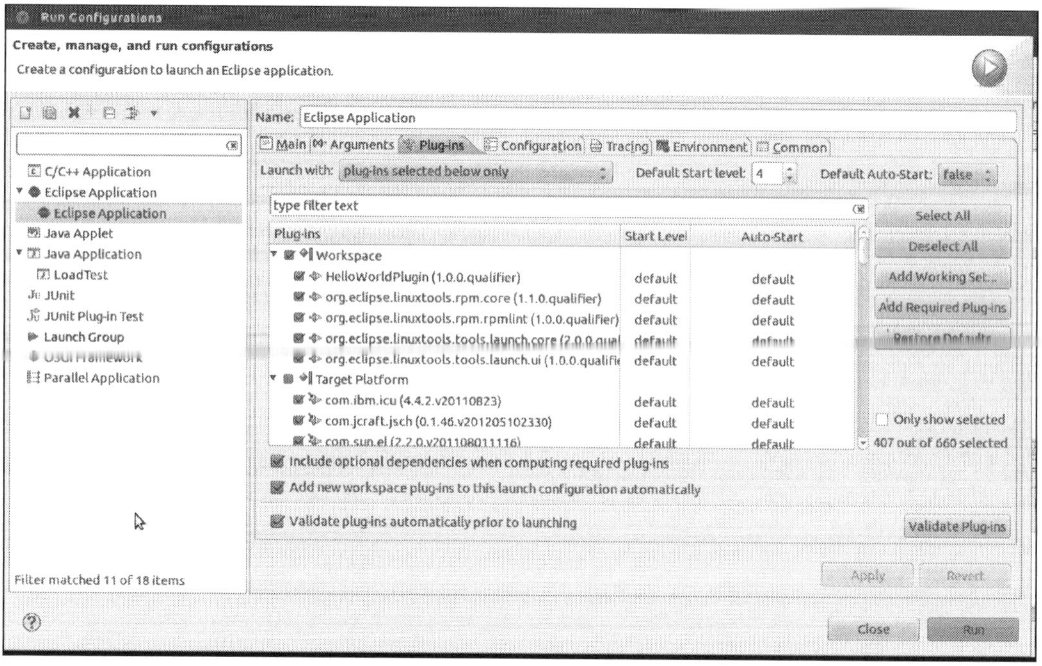

In the **Configuration** tab, you can set the `config.ini` file that you want to use for the runtime workbench instance, as well as modify the configuration area. The **Tracing** tab allows you to enable tracing for each plugin. Tracing is an interesting feature that allows you to turn on or off debug messages during runtime. This can be particularly useful if you are providing support for an Eclipse-based application installed in the client's environment. You can simply require the client to enable tracing for your plugin instead of sending him or her a debug version of the plugin with debug messages enabled. We won't cover tracing on this book. The **Environment** and **Common** tabs are just the same as in a regular Java application.

Extension points

Extension points are the mechanism that Eclipse uses to allow the plugins to extend and modify other plugins' behavior. The main goal of the extension point concept is to provide a loosely-coupled architecture, exposing a few interfaces that can be implemented by other plugins. Both the extension points and its implementations are declared in the `plugin.xml` file.

The Eclipse platform provides a number of these extension points to allow plugins to contribute to various elements of the IDE. We will take a look at some of these extension points throughout this chapter, but a complete list of them can be found in the eclipse documentation, which can be found at `http://help.eclipse.org/indigo/topic/org.eclipse.platform.doc.isv/reference/extension-points/index.html`.

Our Hello World plugin implements one of these extension points. Let's see how it looks like by opening the **Extensions** tab in the plugin manifest editor.

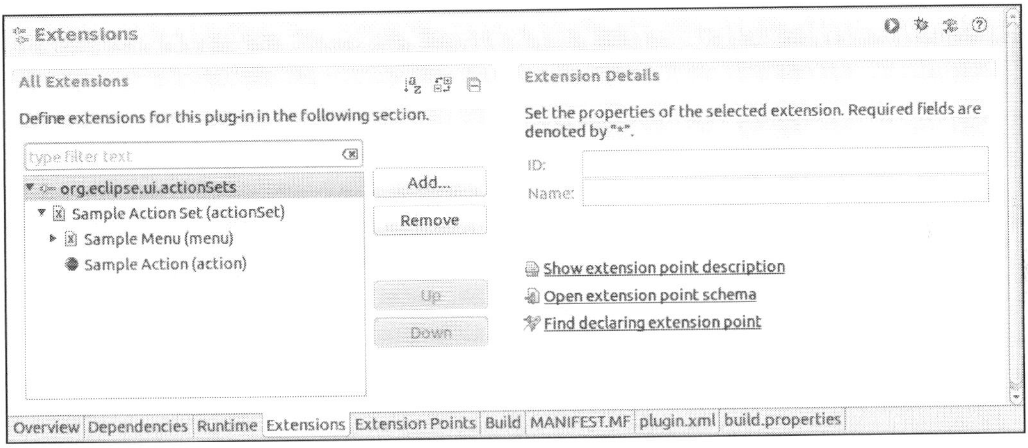

The **Extensions** tab shows us that, in order to add a button to the toolbar or a new entry to a menu, our plugin implements an extension point called **org.eclipse.ui.actionSets**. You can have more information about this extension point by selecting it in the extensions list and clicking on **Show extension point description**.

Declaring an extension point

Currently, our HelloWorldPlugin project only implements extension points provided by other plugins, but doesn't declare any extension points itself. Let's learn how to do it by expanding our Hello World plugin to declare one. We will create an extension point that allows other plugins to provide the message displayed when clicking on the toolbar button.

To create a new extension point, select the **Extension Points** tab in the plugin manifest editor and click on the **Add** button in it. You will be prompted to provide three properties: **ID**, **Name**, and **Schema**. The ID is an internal unique identifier of your extension point. To guarantee its uniqueness, you might want to use the reverse domain name notation to create the ID, containing the domains from less to more specific to which your extension point belongs. Let's use `com.packt.messageProvider` as ID. As you fill this property, the **Schema** is automatically filled with the same content, followed by the `.exsd` extension. The **Point Name** property is a human-readable name for the extension point. Let's use `Message provider` here. When you're done, click on **Finish**. A new editor called schema editor will be opened. The following screenshot shows an example of schema editor:

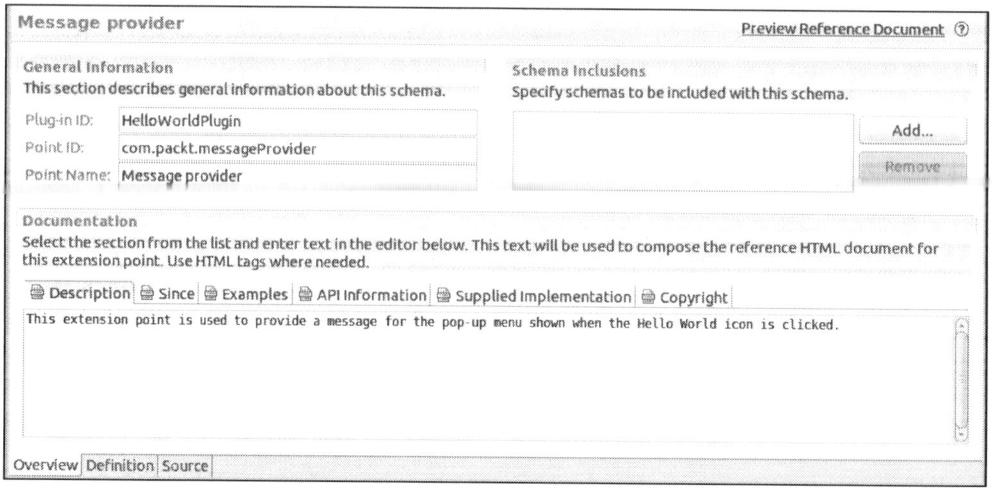

Schemas are XML files that describes extension points. These files are identified by a `.exsd` extension and, just like we did with `plugin.xml`, `build.properties`, and `MANIFEST.MF` files, we will use an editor to add content to it instead of editing it directly. The schema editor contains three horizontal tabs: **Overview**, **Definition**, and **Source**.

The first horizontal tab contains a basic description of your plugin. It also allows you to add documentation to your extension point, which will be shown directly in the Eclipse interface, making it easy for other developers to find information about how to implement your extension point. Remember the `org.eclipse.ui.actionSets` description that we saw in the **Extensions** tab? That's where it comes from. It's very important to thoroughly describe your extension point, as well as provide usage examples and API information.

Now, we have to describe what the extension point contains, that is, what other plugins can provide to our plugin. For this, we have to add new elements to the extension point. Open the **Definition** tab of the schema editor to see a list of the extension point's elements.

Click on **New Element** and change the new element's name to `hello_message`. You can add a small description to the element. Now add an attribute to this element by selecting it and clicking on **New Attribute**. Rename it to `value`. The other properties of the attribute allow you to choose the type that must be provided (string, Boolean, resource, Java classes, or identifiers). In our case, we want the extensions to provide a string. Your **Definition** tab should look as follows:

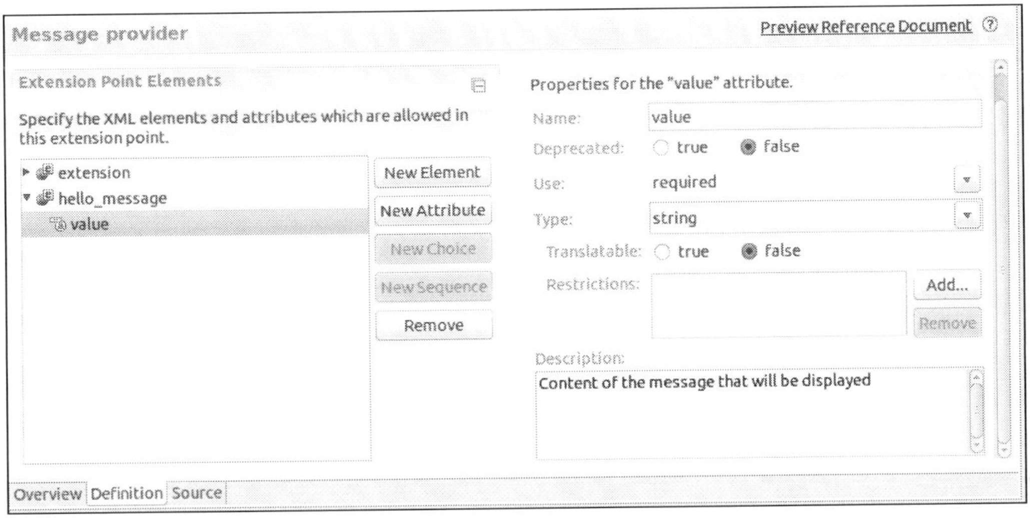

Once the element is defined, we must declare how it composes the extension point. If we had another type of message (a goodbye_message, for example), we could require that the extension point implementation had one or more of each. Element compositions are achieved by adding choices and sequences to the "extension" element. Choices are an "or" composition, while sequences are an "and" composition. These elements can be used together to create more sophisticated compositions.

Since we just want one `hello_message` element per extension point implementation, let's add a new sequence to the extensions element by selecting it and clicking on **New Sequence**. Configure both **Min Occurrences** and **Max Occurrences** to be **1**. Now, right-click on the **Sequence** entry, and select **new | hello_message** to declare that we need a `hello_message` element implementation.

Choices work in a very similar fashion. Add a choice to extension element and add elements to this choice. If any of the elements inside the choice are satisfied, the extension point is successfully implemented.

Providing interfaces

One of the most utilized mechanisms of extension points is to provide interfaces that can be implemented by other plugins. To give an example of how this works, let's modify our extension point a little: instead of requiring a string from other plugins, we'll ask for a class that implements a IMessageProvider interface.

Let's start by creating the interface in our plugin. The interface must be placed inside a package that will be exported in order to be implemented by other plugins, so make sure you create a new package to store the interface. Create a new interface in the project's `src` folder called `IMessageProvider.java` with the following content:

```
public interface IMessageProvider {

    public String getHelloMessage();

}
```

As you can see, the interface only requires a `getHelloMessage()` method that returns a string to be implemented.

Now let's modify the extension point's schema to require an interface implementation instead of a string. To begin with, the name "value" doesn't make much sense anymore, so let's change it to `message_provider`. Now change the **Type** property from string to Java. Two new properties will appear: Extends and Implements. Since we are providing an interface, let's fill the **Implements** field (remember that classes are extended and interfaces are implemented). Click on the **Browse** button and select our IMessageProvider. Remember that you can use the filter to easily find it. Click on **OK**.

The following screenshot shows how our extension point will look in the end:

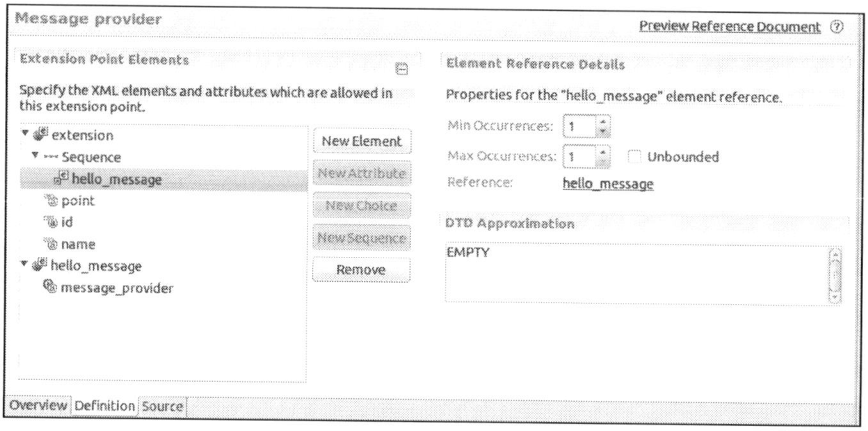

Plugins do not externalize source packages by default. This means that the interface will not be visible to other plugins, thus making it impossible to be implemented. To export the package that contains the interface, we will switch to the **Runtime** tab of the plugin manifest editor. Click on the **Add** button under the **Exported Packages** section, select the package that contains the interface, and click on **OK**.

Using extension points implementation in code

Once the extension point is declared, code in our plugin must be modified in order to utilize what other plugins are providing through it. In our example plugin, this means switching a hardcoded Hello World message for a message provided by an extension point implementation.

Let's take a look at where in the code the Hello World message is actually being displayed. The following piece of code is in the `SampleAction` class, which has been generated by the Hello World plugin project template:

```
public void run(IAction action) {
        MessageDialog.openInformation(
                window.getShell(),
                "HelloWorldPlugin",
                "Hello, Eclipse world");
}
```

This is where we'll work. The first step of utilizing the content coming from the extension points is verifying if there are extensions available and, if there's more than one, which extension the plugin should use. Keep in mind that the whole idea of extension points is to decouple plugins, so your plugin should not crash if there are no implementations to your extension point.

A registry of the available extension point implementations is provided by the Eclipse platform. Once you have it, your code must verify for implementations for your specific extension point, and then look for the elements you need.

This is an example of how the run method should be refactored in order to utilize the message coming from the extension point:.

```
public void run(IAction action) {

    String helloMessage = null;
    IExtensionRegistry extensionRegistry = Platform.
getExtensionRegistry();
    IConfigurationElement[] elements = extensionRegistry.getConfigurat
ionElementsFor("com.packt.messageProvider");
    try{
          for (IConfigurationElement element : elements) {
                Object object = element.createExecutableExtension("mes
sage_provider");
                if (object instanceof IMessageProvider){
                      IMessageProvider messageProvider =
(IMessageProvider)object;
                      helloMessage = messageProvider.
getHelloMessage();
                }
          }
    } catch (CoreException e){
          e.printStackTrace();
    }

    if(helloMessage == null){
          helloMessage = "No providers found";
    }

    System.out.println(helloMessage);
    MessageDialog.openInformation(
          window.getShell(),
          "HelloWorldPlugin",
          helloMessage);
}
```

In the preceding code, we use the `getExtensionRegistry` static method from the `Platform` class to get an `IExtensionRegistry` object that contains all the information about extension points available in the platform. We then use this object to get all the configuration elements available for the `com.packt.messageProvider` extension point. Note that there's no real concern on which extension should be used in case there's more than one it will simply iterate through every configuration element available and assign it to the `helloMessage` variable. There's a fallback in the case where no extensions are found and we get a printed message "No providers found". So regardless of the number of implementations available to our extension point, our plugin code will execute just fine.

Implementing an extension point

Now that we have created the extension point, let's create a new plugin that implements it. Start by creating a new plugin project named `MessageProvider`. We don't want to use any templates this time, so unmark the **Create a plugin using one of the templates** checkbox in the **Templates** page of the new plugin project wizard.

Before we start working with the extension point, we have to declare that the `MessageProvider` plugin depends on `HelloWorldPlugin`, If there's no `HelloWorldPlugin` in the platform, there's no extension point to implement and `MessageProvider` plugin is pointless. To declare the dependency, let's open the plugin manifest editor for the `MessageProvider` project and switch to the **Dependencies** tab. Click on the **Add** button in the **Required Plug-ins** section and type `HelloWorldPlugin`. The following screenshot shows how the **Dependencies** tab should look like:

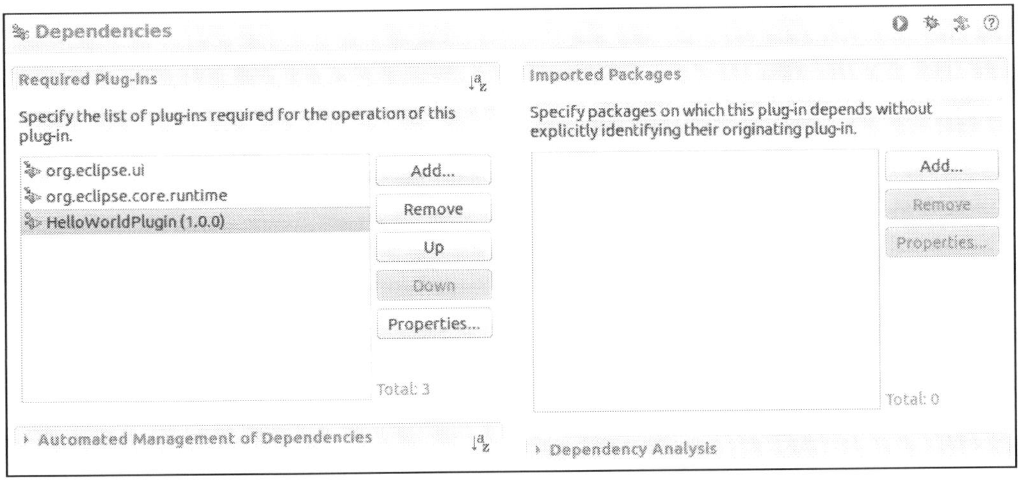

Now we have to declare that our plugin contains an implementation of the `messageProvider` extension point. Switch to the **Extensions** tab, and click on the **Add** button. Select the `com.packt.messageProvider` extension point.

The next step is to actually implement the `messageProvider` extension point. As you can see, the `com.packt.messageProvider` extension already contains a `hello_message` element. The **Extension Element Details** pane shows us what we already know. We need a class that implements IMessageProvider to implement this extension point.

You can create the class by simply clicking on the `message_provider*` link in the `Extension Elements Details` pane. Now copy the following code into it:

```
import helloworldplugin.IMessageProvider;

public class MyMessageProvider implements IMessageProvider {

    @Override
    public String getHelloMessage() {
          return "Hello world from MessageProvider";
    }

}
```

As you can see, this is a very basic implementation. We are now ready to test our extension point and its implementation. Select one of the two projects that we've just created and select **Run | Run as | Eclipse Application**. The message coming from the extension point is displayed when we click on the button, as shown in the following screenshot:

Contributing to the platform's menus and toolbars

In the example that we used to illustrate how to declare and implement an extension point, we created a plugin that adds a button to the Eclipse toolbar. Now, let's take a more careful look at how this button was added there, as well as find out how we could add elements to other parts of the Eclipse interface.

Actions versus commands

The first thing we should know is that there are two different ways of contributing to menus in the Eclipse UI: actions and commands. The action way that we used in the previous example might look simpler at first, but simplicity has a price here; as you can see, logic and interface are closely tied, which might lead to code replication if you want to add, for example, a right-click menu entry that performs the same action as the toolbar button. As you know, code replication is a project's nightmare and should be avoided at all costs. For this reason, usage of actions is discouraged in detriment of commands.

The Command framework solves this issue by completely decoupling handling from interface. It requires implementations for three extension points: `org.eclipse.ui.menus`, `org.eclipse.ui.commands`, and `org.eclipse.ui.handlers`. The menus' extension point is responsible for adding the interface element to the platform. The commands extension point declare commands, which are implemented by handlers.

To show how to use the commands framework, let's implement the same hello world toolbar button that we've just implemented, but now using commands instead of actions. Create a new HelloWorldCommand plugin project without using any templates to start.

org.eclipse.ui.menus

As we already discussed, this extension point is used to add user-interface contributions to Eclipse, and it replaces a number of extension points from the action framework, such as `org.eclipse.ui.ActionSets`, `org.eclipse.ui.EditorActions`, and `org.eclipse.ui.popupMenus`.

Add an implementation of this extension point and a `menuContribution` element with the following parameters:

- **locationURI**: `toolbar:org.eclipse.ui.workbench.file?after=additions`
- **allPopups**: `false`

LocationURI specifies where in the Eclipse UI the contribution will be added. In this case, it will appear right next to the **Print** button. In the next section, we'll learn how to discover locationURIs using the plug-in spy.

If the allPopups Boolean value is set to false, this menu contribution will only be added to context menus that include a marker called "additions".

org.eclipse.ui.commands

The commands extension point declares commands that will be tied to different menu entries. Add an extension to the command extension point and a command element inside it with the following properties:

- **id**: `com.packt.helloworldcommand.command`
- **name**: `helloWorldCommand`

org.eclipse.ui.handlers

This is where the command is actually implemented. This is done by creating a class that extends `org.eclipse.core.commands.AbstractHandler`. So, before adding the implementation for the handlers extension point, let's create a class called `DefaultHelloWorldCommandHandler` with the following code:

```
package com.packt.helloworldcommand;

import org.eclipse.core.commands.AbstractHandler;
import org.eclipse.core.commands.ExecutionEvent;
import org.eclipse.core.commands.ExecutionException;
import org.eclipse.jface.dialogs.MessageDialog;
import org.eclipse.swt.widgets.Shell;
import org.eclipse.ui.PlatformUI;

public class DefaultHelloWorldCommandHandler extends AbstractHandler {

    @Override
    public Object execute(ExecutionEvent event) throws
ExecutionException {
        Shell shell = PlatformUI.getWorkbench().
getActiveWorkbenchWindow().getShell();
        MessageDialog.openInformation(shell, "Hello", "Hello World!");
          return null;
    }

}
```

In the preceding code, the only method you have to implement is `execute`, which is obviously the one that will be called when the command is executed.

Now that we have the handler implemented, let's reference to it in the **Extensions** tab. Create an implementation for `org.eclipse.ui.handlers` and add a `handler` element containing the following properties:

- **commandId**: `com.packt.helloworldcommand.command`
- **class**: `com.packt.helloworldcommand.DefaultHelloWorldCommandHandler`

We have implementations for the three extension points, but note that there's nothing tying up the menu entry with the commands or handlers. This is achieved by adding a "command" element to the `menuContribution` element in the `org.eclipse.ui.menus` extension points. Create one with these parameters:

- **commandId**: `com.packt.helloworldcommand.command`
- **label**: `click me`

Your **Extensions** tab should look like the following screenshot:

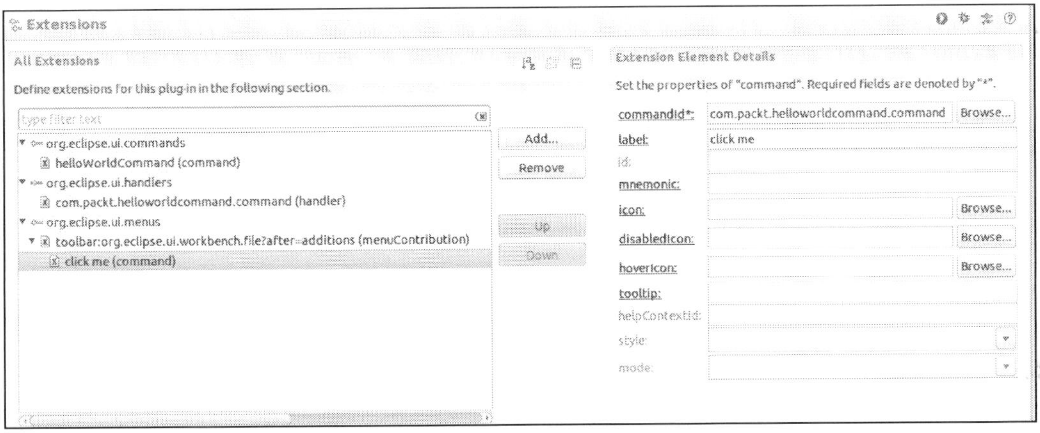

Now let's see how our plugin behaves by running it. You should see a button with **click me** written on it right beside the **Print** button. Click on it to be greeted with yet another hello world message. If you want an icon instead of a text in the toolbar, navigate to **org.eclipse.ui.menus** | **menuContribution** | **command element** to edit it and change the icon parameter.

Since the whole point of the command framework is to allow adding the same command to multiple UI contributions easily, let's test this flexibility by adding an entry to the menu that pops up when the right-mouse button is pressed. All we really need is another `menuContribution` element in `org.eclipse.ui.menus`. So go ahead and create another one, full with the command child element, changing only the `locationURI` property to `popup:org.eclipse.ui.popup.any?after=additions`.

Run the plugin again, and right-click on the Package Explorer in the runtime workbench UI. You should see an entry as follows:

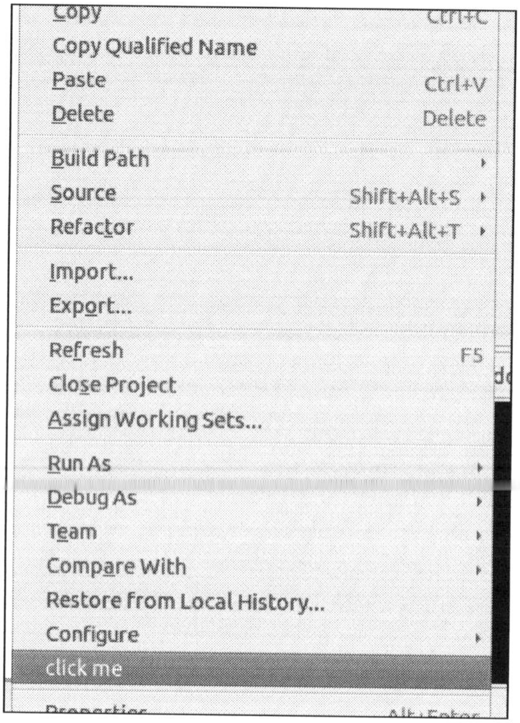

Clicking on it has the same result as clicking the toolbar button.

Restricting and disabling contributions

You might have noticed that the **click me** entry in the right-click menus is not restricted to the Package Explorer. Right-click other views, such as Java Editor or Outline, and you will see that the entry is everywhere. This is probably not the desired behavior; most of the menu contributions apply to some context only, and leaving the entry visible or enabled where they don't apply will only add clutter to the Eclipse interface.

Fortunately, there's an easy way of hiding the menu contribution. The `command` element under `menuContribution` haves an optional element called `visibleWhen`. As the name implies, it allows you to determine under which conditions that command will be enabled.

Let's restrict the right-click menu entry visibility to appear only in text editors. There are two ways of doing this. One is implementing the isEnabled() method in the AbstractHandler extension registered for this command (DefaultHelloWorldCommandHandler in this case). The return value of the method's execution will determine if the contribution is visible or not. The other way is by adding child elements to the visibleWhen element in order to build an expression that is to be evaluated to determine the element's visibility. We'll go for the second option in this example.

Start by adding the visibleWhen element. The checkEnabled property in the element determines which of the two methods of restricting visibility will be used. If true, isEnabled() method will be evaluated, and any child element of visibleWhen will be ignored. If false, child elements will be used. Set this property to false.

There are many types of valid child elements to the visibleWhen element. The with element is utilized to evaluate a variable. These variables are provided by the extension points and the Eclipse framework. A comprehensive list of the variables provided by Eclipse can be found at the Eclipse wiki page in this link: http://wiki.eclipse.org/Command_Core_Expressions#Variables_and_the_Command_Framework. Some examples of such variables are activeEditor, which contains the currently active editor, activeWorkbenchWindow, which contains the currently active workbench window, and activePart, which contains the currently active part. A part is a visual component within the workbench that can be either a view or an editor. Since we want our contribution to be visible in editors only, we'll evaluate the activePart variable. So go ahead and add a with element, and change it's variable value to activePart.

We can add now other elements under the with element to evaluate the activePart variable. activePart contains the instance of the the current active part in the workbench, so to verify if it is a text editor or not, we have to check which classes this object instantiates. This is done with the instanceof element. Let's add it and provide org.eclipse.ui.texteditor.AbstractTextEditor to the value parameter. For this to work, however, we must have access to the org.eclipse.ui.texteditor package, which belongs to the org.eclipse.ui.editors plug-in. Switch to the **Dependencies** tab and add the plugin to the dependencies list. Now switch back to the **Extensions** tab and you should be able to find the AbstractTextEditor class by clicking on the **Browse** button.

Notice that there are many other child elements to visibleWhen, such as "and" and "or", which allow you to compose different conditions. The systemTest element can be used to test system properties, such as osgi.os, that contains the operating system on which Eclipse is running (Win32, Linux, AIX, Solaris, and so on), osgi.ws, which contains the current windowing system (Win32, Motif, GTK, and so on) and osgi.arch, which contains the architecture on which the platform is running (x86, x86_64, PPC, SPARC, and so on).

The following screenshot shows how your `org.eclipse.ui.menus` extension should look:

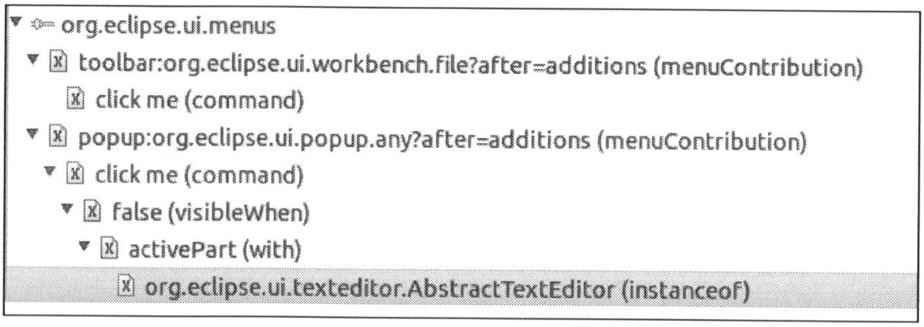

Now, run your project, and try right-clicking on the runtime workbench views. Only text editors will show the **click me** entry.

Menu entries can also be disabled instead of hidden. Take the **copy** entry in a text editor as an example; it's grayed out and un-clickable when there's no text selected, although it behaves normally when there's text selected. Let's add that same behavior to our menu entry to exemplify how it's done.

Entries will be disabled when there's no handler registered for the current situation. Since we haven't conditioned our handler to anything, it's always enabled. Similar to the visibility of `menuContribution`, there are two ways of limiting a handler's activation. You can do it by adding the `enabledWhen` child element to the handler element, and using the same child elements as the ones used in `visibleWhen` (`with`, `and`, `or`, `instanceof`, and so on), assembling an expression that will be evaluated to determine if the handler is active or not. You can alternatively implement `isEnabled()`, returning true or false accordingly. Since we have already shown an example of how to do it using the child elements in the **Extensions** tab, let's override the `isEnabled()` method this time.

The following code is an example of how to override `isEnabled()` in order to restrict the handler to when there's code selected in a Text Editor:

```
@Override
    public boolean isEnabled() {
        boolean enabled = true;
        IWorkbenchPart activePart = PlatformUI.getWorkbench().
getActiveWorkbenchWindow().getPartService().getActivePart();
        if(activePart instanceof IEditorPart){
                ISelection selection = PlatformUI.getWorkbench().
getActiveWorkbenchWindow().getSelectionService().getSelection();
                if(selection instanceof ITextSelection){
```

```
                    if (((ITextSelection) selection).getText().
                            isEmpty()){
                        enabled = false;
                    }
                }
            }
            return enabled;
        }
```

The `PlatformUI` class is the entry point to the Eclipse user interface. We use its static method `getWorkbench()`, and then the `getActiveWorkbenchWindow()` method from the workbench object to get access to `PartService`, which allows us to get the currently active part in the workbench. If the `activePart` variable happens to be an instance of IEditorPart, we get its selection service, that gives us access to what is currently selected in the editor. If the selection happens to be an instance of `ITextSelection`, we verify its content. If the selection is empty, the enabled value is set to false. Note that `ITextSelection` belongs to the `org.eclipse.jface.text` plugin, so this code won't work if this plugin is not specified in the required plugins section from the **Dependencies** tab.

Add this code to the `DefaultHelloWorldCommand` class and run the project. The following screenshot shows the popup menu that you should get if you right-click your Java editor with no code selected:

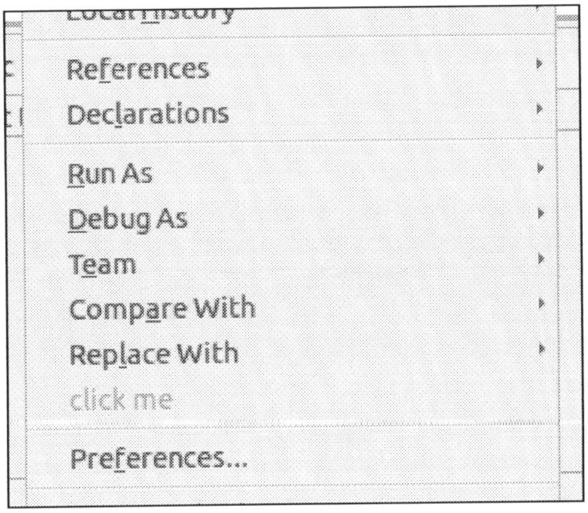

The Plugin Spy feature

You might have wondered where did the locationURIs we used for our contributions came from. As we promised, we'll now show how to use the Plugin Spy, a very useful feature for Eclipse developers, to find locationURIs and much more.

This feature allows you to obtain information from the UI elements of your workbench straight from the Eclipse interface, thus making it easy to find out from which plugin does that element come, its class, the current selection's type, among other useful information. The Plugin spy has two modes: the Plugin Selection Spy and the Plugin Menu Spy. As the names imply, the Selection Spy allows you to get information regarding the currently selected element and its containing part, but the Menu Spy provides information about menu contributions.

To enable the Selection Spy, press *Alt* + *Shift* + *F1*. You will be presented with a pop-up window containing information similar to this one:

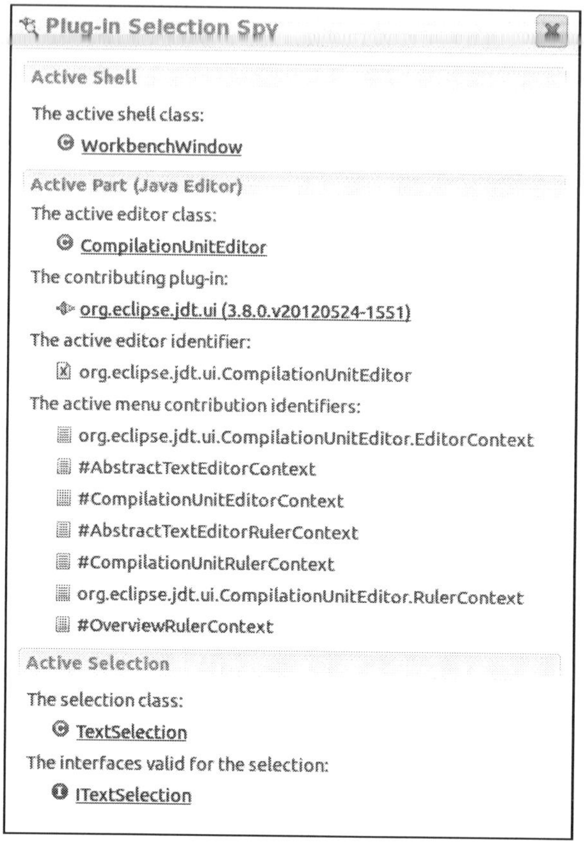

The preceding screenshot is the output of the **Plug-in Selection Spy** with **Java Editor** as the **Active Part**. As you can see, it contains the class of the active editor, the plugin that provides this part, the part's identifier, and the menu contribution identifiers. In the previous example of how to disable menu entries, we have implemented the `isEnabled()` method using information gathered from the plugin selection spy: We know that the Java editor class is `CompilationUnitEditor` and the text selection is an instance of `TextSelection`. If we didn't have the plugin selection spy, we would have to scavenge for this information in the Java Development Tool's code or documentation.

The other Plugin Spy mode is the Plugin Menu Spy. It allows you to get info from menu entries instead of views and selections. To enable this mode, press *Alt* + *Shift* + *F2*. You will notice that your mouse cursor will have a different look. Let's click on in the **Run** toolbar button, for example, to see how the Plugin Menu Spy output looks as follows:

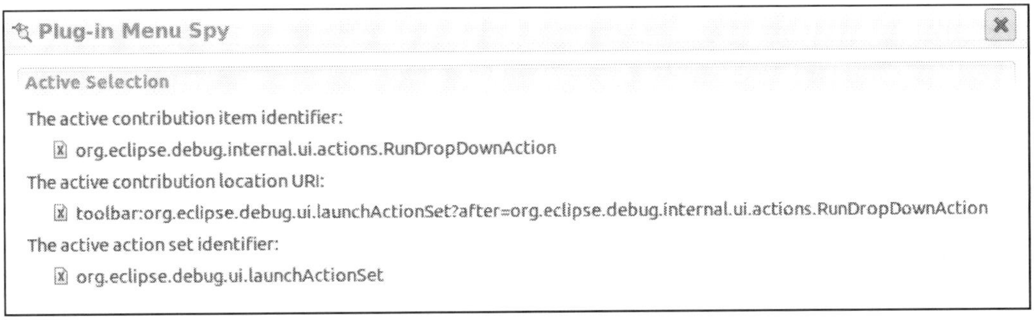

As you can see, it gives us the active contribution item identifier, the location URI of the contribution, and the action set identifier. This field can either point to an action or a command, since the action framework, as we have discussed previously in this chapter, is deprecated but there's still a lot of code that hasn't been ported to the command framework.

The Plugin Menu Spy can also be used to get info from pop-up menus, such as right-click context menus and toolbar drop-down menus.

So now you know from where did we get the locationURI to use in the **click me** menu entry example. Since the Plugin Menu Spy also gives us the action or the command that actually implements the menu entry, it allows us to re-use them by adding new menu contributions that use these actions or commands when required.

To exit the Plugin Menu Spy mode, press *Esc*.

Creating new views

When creating an Eclipse plugin, you might need to create a whole new view instead of extending the current ones. Like almost everything in the Eclipse platform, this is achieved by implementing extension points.

The org.eclipse.ui.views is the most important extension point regarding views. There are three elements in this extension point:

- **View**: Describes the view itself. It contains information regarding the view's class, category, icon, and name
- **Category**: If the view you're creating doesn't belong to any of the existing categories , you can create a new one using this element. You can see all the existing views and its categories by selecting **Window | Show View | Other**.
- **StickyView**: Use this element if you want to create a view that will be shown in all perspectives by default.

The class required by the view element must be one that implements the `org.eclipse.ui.part.IViewPart` interface. It's more recommended, however, to extend the `ViewPart` abstract class, which is a base implementation of `IViewPart` that makes the task of implementing `IViewPart` easier. It leaves two methods to be implemented:

- `createPartControl(Composite parent)`: This is the method that's called when the view is created. According to the method comments of `IViewPart`, the following actions must be taken in implementations:
    - Create one or more controls within the parent
    - Set the parent layout as needed
    - Register any global actions with the site's IActionBars
    - Register any context menus with the site
    - Register a selection provider with the site, to make it available to the workbench's ISelectionService (optional)
- `setFocus()`: This method is called when focus is set in your view. Code in this method should focus one of the elements inside the view.

Basically, every SWT element presented in *Chapter 5, SWT*, and *Chapter 6, More SWT*, can be used to compose a view, as well as your own custom composites.

Saving the view's current state

It's important for most of the views to be able to save its state so that it can be retrieved when Eclipse is restarted. There is a `saveState(IMemento memento)` method in `IViewPart` that must be implemented in order to achieve that. If you take a look at ViewPart's implementation, though, you will be disappointed:

```
public void saveState(IMemento memento) {
    // do nothing
}
```

If you want your view to be capable of saving its own state, you have to override this method. Let's see how to do this.

We can see that the method receives a `IMemento` object as a parameter. This is the object we have to update in order to save the view's state. From the IMemento's JavaDoc, mementos are objects to which you can assign a "mapping of arbitrary string keys to primitive values, and by allowing mementos to have other mementos as children (arranged into a tree)". This mapping is created by calling the `put<type>(String key, <type> value)` methods, such as `putFloat`, `putInteger`, `putString`, and so on. Child mementos are created with the `createChild` method.

Once you have accordingly implemented the `saveState` method, Eclipse guarantees that the information stored in the memento object will be available across different Eclipse sessions, using an xml-based storage format under the hood. However, we still have to tell Eclipse how to use this information to recompose the view's state. This is done by overriding the `init(IViewSite site, IMemento memento)` method, since the ViewPart implementation simply ignores the memento object and delegates to `init(IViewSite site)`. As a general rule, information from the memento is only saved in the current object, leaving the task of assigning the values back to the SWT elements to the `createPartControl` method. This happens because the platform calls `init` before `createPartConrol`.

Adding context help to your view

The easiest way of getting help in Eclipse is through the Dynamic Help, which can be activated in **Help | Dynamic Help**. This opens a **Help** view which contains basic information and usage guidance for the currently focused view. Each view provides the help content, and the Eclipse platform takes care of assembling the help view.

To provide help content, plugins must implement the `org.eclipse.help.contexts` extension point. The implementation must provide a contexts element, which points to a `Context Help` file in the XML format. Instead of creating it from scratch and editing manually, you can use the **Context Help** editor. To do so, create the XML file by selecting **File | New | Other,** and then browsing to **User Assistance | Context Help**. Provide the filename and location in the next page, and click on **Finish**. You can add topics and commands to the context. Topics point to HTML files that will be rendered in the **Dynamic Help** view.

Example of a new view

To illustrate what we have seen about creating a new view so far, let's create a very simple one containing just a label and a text field. Let's begin by extending ViewPart:

```
package sampleview.views;

import org.eclipse.swt.SWT;
import org.eclipse.swt.layout.GridData;
import org.eclipse.swt.layout.GridLayout;
import org.eclipse.swt.widgets.Composite;
import org.eclipse.swt.widgets.Label;
import org.eclipse.swt.widgets.Text;
import org.eclipse.ui.IMemento;
import org.eclipse.ui.IViewSite;
import org.eclipse.ui.PartInitException;
import org.eclipse.ui.part.ViewPart;

public class SampleView extends ViewPart {

    public static final String ID = "sampleview.views.SampleView";

    private Label label;
    private Text text;
    private IMemento memento;

    public SampleView() {
    }

    public void createPartControl(Composite parent) {
            parent.setLayout(new GridLayout(2, false));
                label = new Label(parent, SWT.NONE);
                label.setText("Write down something here:");

                text = new Text(parent, SWT.BORDER);
```

```
                text.setLayoutData(new GridData(SWT.FILL, SWT.NONE,
                                    true, false));
                if(memento != null){
                      if(memento.getString("content") != null){
                       text.setText(memento.getString("content"));
                       }
                }
        }

        public void setFocus() {
                text.setFocus();
        }

        @Override
        public void init(IViewSite site, IMemento memento) throws
                          PartInitException {
            this.memento = memento;
            init(site);
        }

        @Override
        public void saveState(IMemento memento) {
            memento.putString("content", text.getText());
        }
    }
```

Now that we have the `IViewPart` implementation, let's modify `plugin.xml` in order to register the view in the Eclipse platform. The following screenshot shows how our **Extensions** tab of the plugin manifest editor looks like after registering the view:

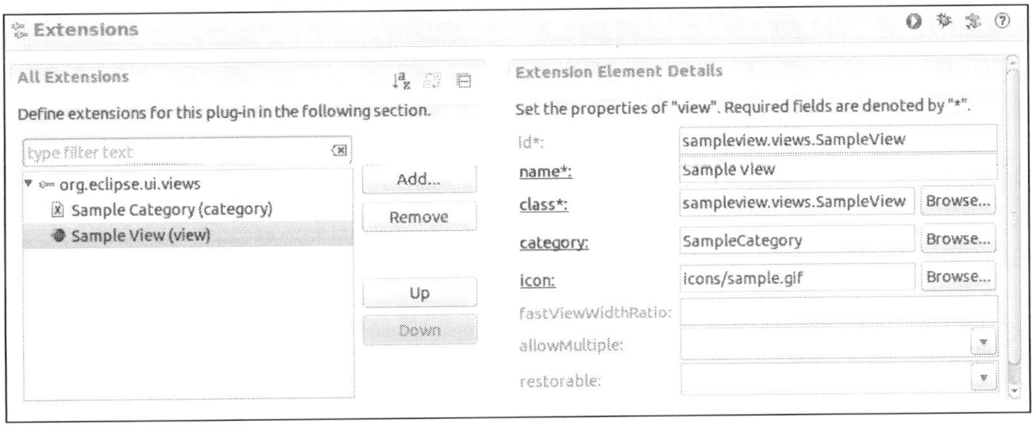

Now run the project. Open the view that we just created by selecting **Window | Show View | Other** or by using the *Ctrl + 3* shortcut and typing the view's name. Our view should look like this:

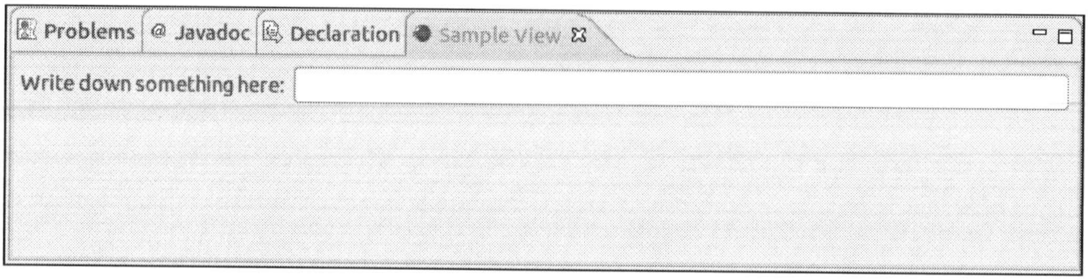

Try writing something in the text field and restarting the runtime workbench. Note that you must close the Eclipse instance instead of clicking the **Terminate** button (red stop button) in your development environment. The **Terminate** button simply kills the instance, giving no time for the platform to save your view's state.

Exporting a plugin

So far we have only executed our plugins in our runtime workbench. In order to provide our plugin to be installed in other machines, we have to export it. To do it, right-click your plugin project and select **Export**. The export wizard contains four different ways of exporting plugin projects. We will use the **Deployable plug-ins and fragments** option. Select it and click on **Next**.

The next wizard page allows you to choose the plugins you wish to export. There are three possible destinations for your plugins: a directory, an archive file, or a host. If you choose to export your plugins into a directory, each plugin will be bundled into its own `.jar` file and placed inside a plugins folder. To install your plugins, simply move these jar files into the plugins folder inside your Eclipse installation folder. The archive option does the same thing, but on top of that archives the plugins folder into a `.zip` file. The **Install into host** option allows you to deploy your plugin directly into an Eclipse installation.

The **Options** tab allows you to also export your plugin's source code. If you select this option, the JAR file containing your plugin will also contain a `src` folder with all your classes' code in it. The **JAR Signing** tab lets you sign your plugin's JAR file using a keystore.

Summary

As you have seen in this chapter, Eclipse provides a very good set of tools to developers interested in extending it. The plugin manifest editor makes it easier to start creating plugins by removing the need to know how to edit the plugin's configuration files. The plugins that are provided by Eclipse are very well-documented, and together with features such as the Plugin Spy, will get you going in no time.

Now that we have some basic understanding of how to develop extensions for Eclipse, in the next chapter we will find out how we can use all of the Eclipse's platform and infrastructure to develop our own client applications.

9
Eclipse Rich Client Platform

In the previous chapter, we have talked about plugin development for Eclipse. We also saw that **Eclipse** is composed of a basic platform and plugins that provide all of the functionalities of the **IDE**. What we didn't tell you is that you can use this platform to build your own client applications from scratch, taking advantage of everything Eclipse offers. Imagine Eclipse without any views, perspectives, or toolbars – this is what the **Rich Client Platform** (**RCP**) provides. It is an empty canvas onto which you can develop your application using all of Eclipse's building blocks, and any other ones you might come up with.

In this chapter we will:

- Learn what a Rich Client Platform is and how developers can benefit from building their applications using it
- See what the Eclipse Rich Client Platform is and what it provides
- Develop a very simple client application using **Eclipse RCP** to showcase how to use its most important features
- Learn how to bundle and brand your client application

Understanding a Rich Client Platform

Before plunging into the Eclipse RCP, let's define what a Rich Client Platform is. In the context of client/server architecture, the client application is responsible for getting requests from the user and sending them to a machine that will service the requests, accessing databases and internal systems. The popularization of the Internet and improvements in speed and stability has led many companies to switch their client applications from desktop to Internet-based ones. This switch conveys a number of benefits, such as ease of deployment and updates, because any modification can be deployed to all clients by simply modifying the code of the application in the web server. Therefore, there are less hardware and software requirements for the client side, among others benefits. However, poor user experience due to network latency and downtime caused by connection issues, which even though rare, can be crucial for business, kept alive the interest in desktop-based applications.

Developing one of these client applications from scratch, however, is an expensive and time-consuming task. A number of different aspects must be thought of, such as portability, modularity, and update methods. That's where a Rich Client Platform comes in. It offers all of these features for developers to build their own client applications without having to worry about the different aspects for every type of client application they work on. Since client applications' user interfaces don't differ much from each other, most of them are basically composed of forms with data-input elements, such as text fields and combo boxes. An RCP should also provide reusable widgets that can easily be added to the applications.

Eclipse is not exactly a client application from a client/server architecture perspective. It contains all of these desirable features, and they are made available through the Eclipse RCP for developers interested in taking advantage of them for their own client applications.

The Eclipse Rich Client Platform

Now that we know what a rich client platform is, let's see in more detail what features the Eclipse RCP provides.

OSGi framework implementation

Eclipse RCP provides **Equinox**, an implementation of the **OSGi framework**. The OSGi framework, according to its specification document, provides a general-purpose, secure, managed Java framework that supports the deployment of extensible and downloadable applications known as bundles. Equinox provides to your client application all the infrastructure it needs to be a modularized application that can easily ship and orchestrate pieces of software from various origins.

The functionalities that the OSGi framework specification provides can be divided into different layers. Let's see these in detail.

The module layer

The module layer extends the Java platform to provide packaging, deploying, and validating services to Java-based projects. This layer defines a unit of modularization named bundle, which contains the necessary resources to provide a certain functionality. Once loaded, a bundle can also provide services to other bundles through the OSGi service platform. Bundles translate to plugins in the Equinox implementation.

Bundles are JAR files with a specific format and content. They must contain a MANIFEST file, which determines what the bundle provides to other bundles, the services it provides for the end user, and the resources that must be available in order to load the bundle.

The life cycle layer

The life cycle layer determines the **API** (**Application Programming Interface**) that provides services to manage all the phases of a bundle's life cycle, such as installation, starting, stopping, updating, and uninstallation. Before installation, the bundle is verified for a number of errors, such as compatibility between the bundle and current execution environment and the previous installation of the bundle with the same name and version. If no errors are detected, the bundle is considered valid and can be installed in the OSGi framework. Once installed, the bundle can be started. Before starting a bundle, the OSGi framework must resolve it, which means verifying if all of the bundle's dependencies are met in the current OSGi service platform. The bundle is started by calling its `Activator` object, which is determined in the MANIFEST file. When the bundle is stopped, all of its resources must be released and the services it provides must be unregistered. Uninstalling a package notifies other bundles about the removal of the bundle and removes every resource associated to it from the persistent storage.

The service layer

The service layer defines the model through which bundles share functionalities between themselves. This layer is tightly integrated with the life cycle layer, as it determines how the bundles' state modifications, such as installation, starting, and stopping, should notify other bundles about its services. Bundles can register or unregister services at any time. Once registered, other bundles can search for that service and utilize them.

The security layer

The security layer is an optional layer that adds security services based on the Java 2 specification. In general terms, it specifies how JAR files that represent bundles (plugins) must be signed so that their authenticity can be verified. This includes determining which signing algorithms can be used and where hash files must be placed in the JAR file, among others.

SWT

The Eclipse RCP provides the **Standard Widget Toolkit** (**SWT**) to supply you with all the building blocks you might need to create your application's user interface. This means that every widget presented in *Chapters 5, SWT* and *Chapter 6, More SWT* can be used here.

JFace

JFace is a user interface framework that goes on top of the widget toolkit. Its main objective is to make UI programming easier and increase code reuse by providing the following among others:

- Bigger building blocks than the ones provided by SWT, such as wizards, dialogs, and context help menus
- Image descriptors and registry, lifting the developer from the obligation of loading, and managing and disposing images
- Viewers that help populating, sorting, and updating widgets

JFace does not intend to completely hide SWT. It allows the developer to choose between using JFace or SWT directly, depending on what is more convenient in each situation. While JFace is completely dependent on SWT, SWT does not depend on JFace whatsoever.

We have already used some of the JFace features in *Chapter 8, Eclipse Development*. We utilized the `MessageDialog` class to display the `hello world` message without having to worry about creating the dialog and adding its elements, such as the **OK** button, the information icon, and the message itself. We have also used the `ITextSelection` class to get the text selected in a text editor.

Eclipse workbench

The Eclipse RCP also allows the Eclipse workbench to be utilized in client applications. You can re-use Eclipse's toolbars, views, editors, and perspectives according to your needs, as well as make custom ones specifically for your application, extending existing Eclipse classes, or making them from scratch. Given that, everything we saw in the *Creating New Views* section of the previous chapter is also valid here. You will have the whole underlying platform that's required to display and manage the arrangement of these UI elements.

Developing a client application using the Eclipse RCP

The best way of showing the basics of client application development with the Eclipse RCP is by using an example. As we have seen in *Chapter 8, Eclipse Development*, Eclipse contains useful plugin project templates that can get you going into plugin development in no time. The same goes for RCP applications. Let's create a new RCP application project using a template to have an idea of what a basic RCP application project looks like.

Creating a new client application project using a template

The process of creating an RCP application is very similar to that of creating a plugin project. Navigate to **File | New | Other**, and in the wizard window, select **Plug-in Project**. Click on **Next**. In the next wizard window, fill in the project's **Name** field with something such as `HelloRCP`, and then click on **Next**. In the next window, select **Yes** for the **Would you like to create a rich client application** question, as shown in the following screenshot:

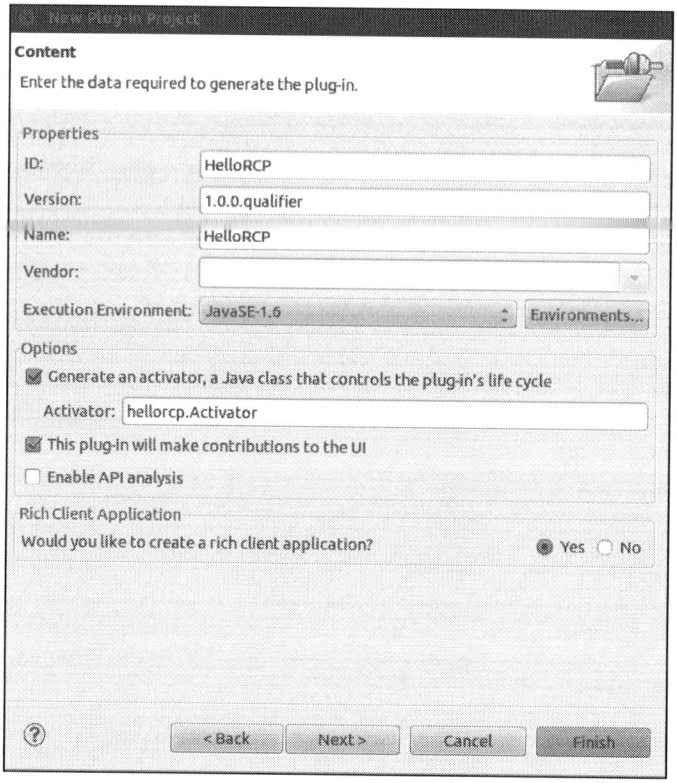

The next wizard window contains the available RCP templates. Let's use the **RCP application with a view** template. The last page contains some basic information required by this template. Let's leave the default values and click on **Finish**. You will be asked if you want to switch to the **Plug-in Development** perspective at this point if you're not already using it. This perspective adds views and menu entries that can be very useful for plugin development, so go ahead and choose **Yes**. The project will appear in the **Package Explorer** window and the **Plug-in Manifest Editor** window will be opened.

If you take a look at the project's contents, you will notice that they are very similar to the contents of a Plug-in project. This is because an RCP application is actually a plugin. The implementation of the `org.eclipse.core.runtime.applications` extension point, which can be verified in the **Extensions** tab of the **Plug-in Manifest Editor** window, sets it apart from a regular plugin project. Notice that the Application element points to a class named `Application`.

A closer look at the generated source code

We will now look at the content of classes generated by the template to understand what they do and why they are there. Let's start with the `Application.java` source file. As you can see, this class is an implementation of the `IApplication` interface, which contains the following two methods:

```
public Object start(IApplicationContext context) {
  Display display = PlatformUI.createDisplay();
  try {
    int returnCode = PlatformUI.createAndRunWorkbench(display, new
ApplicationWorkbenchAdvisor());
    if (returnCode == PlatformUI.RETURN_RESTART) {
      return IApplication.EXIT_RESTART;
    }
    return IApplication.EXIT_OK;
  } finally {
    display.dispose();
  }
}
```

And

```
/* (non-Javadoc)
 * @see org.eclipse.equinox.app.IApplication#stop()
 */
public void stop() {
  if (!PlatformUI.isWorkbenchRunning())
    return;
  final IWorkbench workbench = PlatformUI.getWorkbench();
  final Display display = workbench.getDisplay();
  display.syncExec(new Runnable() {
    public void run() {
      if (!display.isDisposed())
        workbench.close();
    }
  });
}
```

As the names imply, the `start` method is called during the application's startup, and the `stop` method is called when the application is stopped. The `start` method can return any type of object, including integers, as is the case with the `Application` implementation. This is a very basic implementation that simply creates a new `Display` object and calls the static `PlatformUI.createAndRunWorkbench` method. This method receives a `WorkbenchAdvisor` object as a parameter. This object allows you to configure some basic aspects of your application's look and feel. Every RCP application must implement a `WorkbenchAdvisor` object. The `Display` object manages the connection between SWT and the operating system. For most RCP applications, you won't need much more than these in the `start` method to get it going. The `stop` method simply closes the workbench created during the `start` method.

Now let's take a closer look at the generated extension of the `WorkbenchAdvisor` class, the `ApplicationWorkbenchAdvisor` extension:

```
public class ApplicationWorkbenchAdvisor extends WorkbenchAdvisor {

  private static final String PERSPECTIVE_ID = "HelloRCP.perspective";

  public WorkbenchWindowAdvisor createWorkbenchWindowAdvisor(
      IWorkbenchWindowConfigurer configurer) {
    return new ApplicationWorkbenchWindowAdvisor(configurer);
  }

  public String getInitialWindowPerspectiveId() {
    return PERSPECTIVE_ID;
  }

}
```

The `getInitialWindowPerspectiveId` method determines which perspective will be displayed when a new window of our application is opened. The `CreateWorkbenchWindowAdvisor` method is responsible for instantiating and returning a `WorkbenchWindowAdvisor` object, which configures the application window itself. Our implementation returns an instance of `ApplicationWorkbenchWindowAdvisor` extension, which extends `WorkbenchWindowAdvisor` object. Let's take a look at its implementation:

```
public class ApplicationWorkbenchWindowAdvisor extends
WorkbenchWindowAdvisor {

  public ApplicationWorkbenchWindowAdvisor(IWorkbenchWindowConfigurer
configurer) {
    super(configurer);
  }
```

```
    public ActionBarAdvisor createActionBarAdvisor(
        IActionBarConfigurer configurer) {
      return new ApplicationActionBarAdvisor(configurer);
    }

    public void preWindowOpen() {
      IWorkbenchWindowConfigurer configurer = getWindowConfigurer();
      configurer.setInitialSize(new Point(400, 300));
      configurer.setShowCoolBar(false);
      configurer.setShowStatusLine(false);
      configurer.setTitle("RCP Application");
    }
  }
```

The `preWindowOpen` method implementation calls `getWindowConfigurer` and performs some actions on the `IworkbenchWindowConfigurer` configurer, such as setting an initial size for the application window, disabling the coolbar and status line, and setting a title for the application. The `createActionBarAdvisor` method instantiates and returns `ApplicationActionBarAdvisor`, which extends `ActionBarAdvisor`. `ActionBarAdvisor` creates, adds, and disposes the JFace actions of a workbench window. A JFace action is a command that can be triggered by the end user utilizing UI elements, such as buttons, menu items, and items in the toolbar. The `ApplicationActionBarAdvisor` implementation doesn't modify `ActionBarAdvisor`'s behavior.

Now, let's go back to the **Plug-in Manifest Editor** window and see what other plugins are required to build a fully functional RCP application. As we said in the beginning of the chapter, an RCP application is something like an Eclipse instance stripped of every Eclipse graphical element. This means you must create these elements if you want your application to function properly. The first thing we need is a perspective, which is created by implementing the `org.eclipse.ui.perspectives` extension point. The perspective element contains an ID, which must be unique across the applications, must be a user-friendly name that will be displayed in the interface, and a class. Let's take a look at the `Perspective` class code:

```
public class Perspective implements IPerspectiveFactory {

  public void createInitialLayout(IPageLayout layout) {
    layout.setEditorAreaVisible(false);
    layout.setFixed(true);

  }
}
```

The code shows us that perspective classes must implement the `IPerspectiveFactory` interface, which contains the `createInitialLayout` method. The implementation generated by the template simply removes the editor area from the application and disallows users to modify the view's size and placement by calling the `setFixed` method. The editor area is a space in the center of the application's window that is reserved for editors. Any other views will be rendered either below, to the left of, or to the right of this area. The editor area is disabled for our application because it doesn't have any editors.

This method could also be used to add the views to the perspective by calling the `layout.addView` method. However, the code generated by the template uses a different approach. Go back to the **Plug-in Manifest Editor** window and take a look at the `org.eclipse.ui.perspectiveExtensions` implementation. This extension point allows you to add a view to an existing perspective. This approach is generally used when you want to add a view to a perspective that wasn't developed by you. In this case, you wouldn't have access to the `IPerspectiveFactory` implementation to add a `layout.addView` call to the `createInitialLayout` implementation.

The `perspectiveExtension` element is where you define the ID of the perspective that will receive the view. This field allows using the `*` (asterisk) character, which will cause the view to be added to every perspective that matches the pattern. For example, `org.eclipse.jdt.*` would include the view in every **Java Development Tools (JDT)** perspective. This is another benefit of using the `perspectiveExtensions` extension point instead of calling `layout.addView` in the `IPerspectiveFactory` implementation. It simplifies the task of managing the perspectives into which the view is added. In our example, `targetID` is simply `*`, which means that in `HelloRCP.view`, the view will be added to every perspective available (there's only one perspective in our application).

Speaking of the view, let's take a look at the `org.eclipse.ui.views` extension. This is a simple view, just like the one we created in *Chapter 8, Eclipse Development*. It points to a View class that extends the `ViewPart` abstract class. The `createPartControl` method instantiates a `TableViewer` object and associates a content provider and a label provider to it. We will explore the concepts of a label and content providers when we expand this example in the later sections. `TableViewer` is a JFace widget that displays content in a table-like list, similar to the one utilized in JDT's Problems view.

The last extension point that our RCP application implement is `org.eclipse.ui.menus`. It simply provides a **File** menu contribution that contains an **Exit** entry. This entry runs the `org.eclipse.ui.file.exit` command, which causes the program to exit.

Running and debugging RCP applications

The easiest way of running an RCP application is by opening the **Plug-in Manifest Editor** window for the RCP project, switching to the **Overview** tab, and clicking on the **Launch an Eclipse application** link on the column to the right. This will create a default configuration containing the RCP plugin and the necessary platform plugins. You can also create custom run configurations, just like with regular plugins, by selecting your project and navigating to **Run | Run Configurations** in the Eclipse menu. Double-click on the **Eclipse Application** option on the left to create a new run configuration for the selected project. To run your project as an RCP application instead of an Eclipse plugin, make sure you switch to **Run an application** and select your application in the **Program to run** section, as shown in the following screenshot:

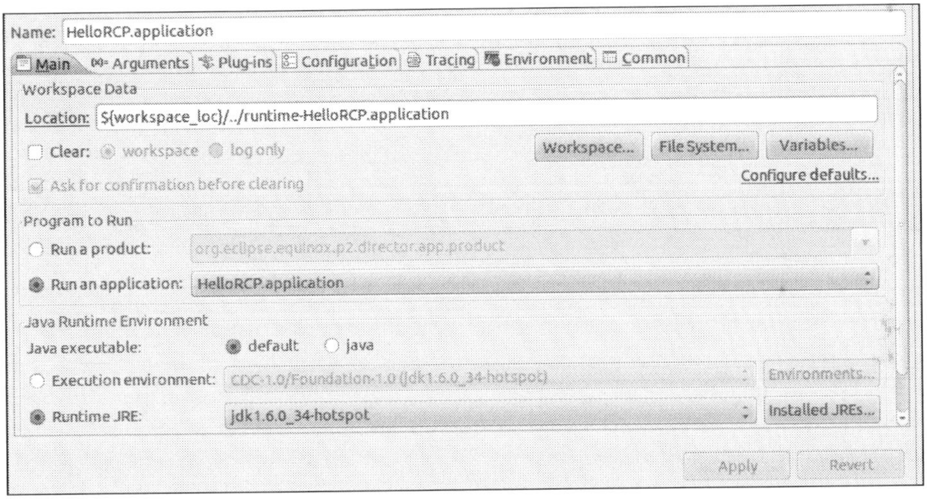

This configuration will also contain all the plugins from your development platform. To select only the required plugins, switch to the **Plug-ins** tab, and select **plug-ins selected below only** from the **Launch with** drop-down menu. Click on **Deselect all**, select your RCP plugin, and click on **Add Required Plug-ins** to have only the necessary platform plugins in your application.

The configuration options for the other tabs presented in *Chapter 8, Eclipse Development*, are also valid here. You can also go to **Run | Run Configuration** and Eclipse will create a default run configuration for you if you haven't created one yet. You will be presented with a number of different ways to run your app: as an Eclipse application, a Java applet or application, or as an OSGi Framework application. Select **Eclipse Application** and click on **OK**. This configuration will contain only the currently selected plugin and the necessary platform plugins.

The following screenshot shows what the sample application we built in the previous section looks like. As you can see, it complies with all our specifications: a 400 x 300 pixel window, a view, the **RCP Application** title, and so on.

Debugging is also a straightforward operation—just place breakpoints in the classes you wish to debug and navigate to **Run** | **Debug**. It will use the same configurations as the **Run** command.

Expanding the example – creating a contact list application

Now that we have seen the basics of an Eclipse RCP application, let's expand the example and create an application to store contact information. To keep it simple and focus on learning how to use the Eclipse RCP features, we will develop a standalone application instead of a client application that connects to a server. The application, however, could be easily modified to store and fetch contact information from a server. The advantage of such a design over a web-based client lies in the possibility of implementing mechanisms that allow your application to work in situations where Internet connection is faulty or simply nonexistent.

The idea here is to develop a very simple application, but one that has functionality and explores more features than the one in the previous example. It will have a view on the left side of the application window, containing a list of the contact names. Whenever a contact name is selected in this view, a contact editor is opened in the middle of the window, allowing the user to view the contact information and update it. The **application** menu will contain entries to create a new contact, save the currently edited one, as well as a toolbar with a **save** button. We will also use this example to introduce the concept of content and label providers.

We will create a number of classes in this example. The following screenshot contains all of them in their respective packages so that you can refer to it whenever you need a big picture of the application:

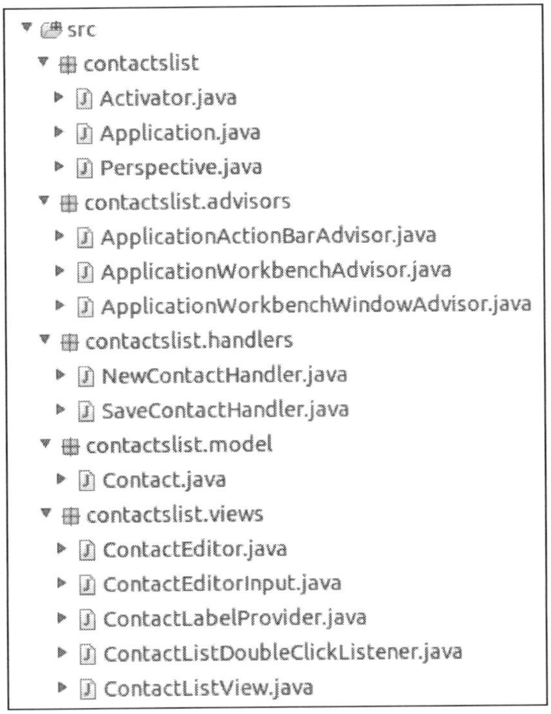

Ok, so let's get our hands on it. As you have seen in the previous example, Eclipse templates are a good way of getting started; so let's use it in this example too. Navigate to **File** | **New** | **Plug-in Project** and create a new RCP project named `ContactsList` using the `HelloRCP` template. Change the application window title to something more appropriate, such as `Contacts List`.

The project has a similar structure to the previous example, but with fewer classes. Let's run the application to make sure everything is okay. Open the **Plug-in Manifest Editor**, switch to the **Overview** tab, and click on the **Launch an Eclipse Application** link in the column to the right. You will see an empty window named `Contacts List`, which is exactly what we expected by now. Let's close it and proceed to developing the application.

The contact list view

We will start by coding the view that contains a list of the contacts. To create this list, we will use the SWT's `TableViewer` element. Even though this is a very basic view, we will use label and content providers for the sake of demonstration. A label provider is a special class that is responsible for mapping models to their corresponding entries in a viewer. In our `TableViewer` case, it will provide a label that represents contact. We will use the contact's name for that. The content provider feeds the viewer with the elements it should contain. We will use a simple SWT implementation named `ArrayContentProvider`, which supplies elements via an array.

Let's begin with the contact model. Create a class named `Contact` that contains three strings along with their getters and setters: `name`, `address`, and `phoneNumber`. Also create a constructor with all the strings.

The next class we will work on is the label provider. Create a class named `ContactLabelProvider` and make it implement the `ILabelProvider` interface. Use the Eclipse code generation features presented in *Chapter 2, Java Development*, to create default implementations for the interface's methods. The only method we will modify is `getText`. This is where we will determine that the contact will be represented by its name in the contact list. We need to get the `element` parameter, check if it is an instance of `Contact`, and return the contact's name. The following code does that:

```
@Override
public String getText(Object element) {
  if(element instanceof Contact){
    return ((Contact)element).getName();
  }
  return null;
}
```

Now that we have everything we need, let's start working on the `ContactList` view itself. Create a class named `ContactListView` with the following code:

```
public class ContactListView extends ViewPart {

  private TableViewer contactsViewer;
  private List<Contact> contacts;
```

```
    @Override
    public void createPartControl(Composite parent) {
      contactsViewer = new TableViewer(parent);
      contactsViewer.setContentProvider(new ArrayContentProvider());
      contactsViewer.setLabelProvider(new ContactLabelProvider());
      contacts = new ArrayList<Contact>();
      contactsViewer.setInput(contacts);
    }

    @Override
    public void setFocus() {

    }

    public void addEntry(Contact contact){
      contacts.add(contact);
      contactsViewer.refresh(true);
    }

    public void refreshEntries(){
      contactsViewer.refresh(true);
    }
  }
```

The imports section was omitted. As we have seen in the previous chapter, an easy way of creating a view is by extending the `ViewPart` class. This requires us to implement the `createPartControl` and `setFocus` methods. In the `createPartControl` method, we are supposed to instantiate all of the SWT elements contained in the view. The `setFocus` method is executed when focus is assigned to the view. This method typically assigns focus to one of the SWT elements in the view. We will ignore `setFocus` for now and use an empty implementation. In the `createPartControl` method, we initialize the `TableViewer` interface and associate it with the `ContactLabelProvider` class we just created. We also create a new `ArrayContentProvider` method and supply it with an empty `ArrayList` instance. This list will be modified as we create new contact entries. We have also created two methods (`addEntry` and `refreshEntries`) to simplify and encapsulate the task of adding new contacts to the list.

Now that we have the `ContactList` class, let's modify the `plugin.xml` file to register it in the workbench. We will use the **Plug-in Manifest Editor** window for that, just like we did in the previous chapter. Double-click on the `plugin.xml` file to open it, switch to the **Extensions** tab, and add a new `org.eclipse.ui.views` entry. Add a new **view** element to the extension. The following screenshot contains the properties of this element:

We must also add our view to the default perspective of our application. Let's use `perspectiveExtension` for that, just like we did in the previous example. Create an extension to `org.eclipse.ui.perspectiveExtensions`, add a `perspectiveExtension` with `*` as `targetID`, and a **view** element under it with the properties as shown in the following screenshot:

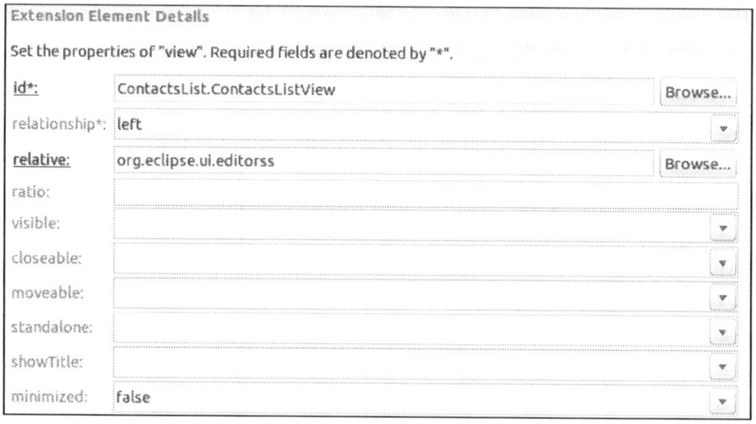

The **relative** field determines which part will be the reference for the positioning of our **view** window. In this case, we're using the editor space (`org.eclipse.ui.editors`), which represents the currently active editor in the workbench. Clicking on the **Browse** button allows searching for parts IDs. The **relationship** field determines where it will be placed relative to the part selected in the **relative** field.

The contact editor

In order to create new contact entries and edit the existing ones, we will create a contact editor. Before we work on the `ContactEditor` class itself, let's create an editor input for it. An editor input represents the entry that is being editing. An editor input for a file, for example, could contain the path for the file being edited. Because our application's contents will be stored in-memory only, the editor input must contain the `Contact` object itself. Create a class named `ContactEditorInput` that implements the `IEditorInput` interface, with the following code:

```
public class ContactEditorInput implements IEditorInput {

private Contact contact;

  public ContactEditorInput(Contact contact) {
    super();
    this.contact = contact;
  }

  public Contact getContact(){
    return contact;
  }

  @Override
  public String getToolTipText() {
    return contact.getName();
  }


  @Override
  public String getName() {
    return contact.getName();
  }

  @Override
  public boolean equals(Object obj) {
    if(obj instanceof ContactEditorInput){
      if(((ContactEditorInput)obj).getContact().equals(this.contact)){
        return true;
      }
    }
    return false;
  }

  //Auto-generated method stub
}
```

We have created a constructor that receives a `Contact` object and also implemented a getter for it. We have also implemented the `getToolTipText` and `getName` methods to return the contact name. The `equals` method must also be implemented in order to make sure that only one editor is open for one contact. We have used the auto-generated method `stub` for the other methods. These methods were omitted in the code.

Now let's work on the contact editor. The following code is the implementation of the `ContactEditor` class:

```
public class ContactEditor extends EditorPart {

  private boolean dirty = false;
  private Text addressText;
  private Text nameText;
  private Text phoneText;

  @Override
  public void init(IEditorSite site, IEditorInput input)
      throws PartInitException {
    setSite(site);
    setInput(input);
  }

  @Override
  public void createPartControl(Composite parent) {
    parent.setLayout(new GridLayout(2, false));
    Label nameLabel = new Label(parent, SWT.NONE);
    nameLabel.setText("Name");
    nameLabel.setLayoutData(new GridData());
    nameText = new Text(parent, SWT.BORDER | SWT.FILL);
    nameText.setLayoutData(new GridData());

    Label addressLabel = new Label(parent, SWT.NONE);
    addressLabel.setText("Address");
    addressLabel.setLayoutData(new GridData());
    addressText = new Text(parent, SWT.BORDER);
    addressText.setLayoutData(new GridData());

    Label phoneLabel = new Label(parent, SWT.NONE);
    phoneLabel.setText("Phone number");
    phoneLabel.setLayoutData(new GridData());
    phoneText = new Text(parent, SWT.BORDER );
    phoneText.setLayoutData(new GridData());
```

```
    ContactEditorInput editorInput = (ContactEditorInput)
getEditorInput();
    if(editorInput != null){
      Contact contact = editorInput.getContact();
      nameText.setText(contact.getName());
      this.setPartName(contact.getName());
      addressText.setText(contact.getAddress());
      phoneText.setText(contact.getPhoneNumber());
    }

    nameText.addModifyListener(new ContactEditorModifyListener());
    addressText.addModifyListener(new ContactEditorModifyListener());
    phoneText.addModifyListener(new ContactEditorModifyListener());
  }

  @Override
  public void doSave(IProgressMonitor monitor) {
    dirty = false;
    firePropertyChange(IWorkbenchPartConstants.PROP_DIRTY);

    ContactEditorInput editorInput = (ContactEditorInput)
getEditorInput();
    if(editorInput != null){
      Contact contact = editorInput.getContact();
      contact.setName(nameText.getText());
      contact.setAddress(addressText.getText());
      contact.setPhoneNumber(phoneText.getText());

      this.setPartName(contact.getName());
    }
  }

  @Override
  public boolean isDirty() {
    return dirty;
  }

  @Override
  public void doSaveAs() {
  }

  @Override
  public boolean isSaveAsAllowed() {
    return false;
```

```
    }

    @Override
    public void setFocus() {
    }

    private class ContactEditorModifyListener implements ModifyListener
  {
      @Override
      public void modifyText(ModifyEvent e) {
        dirty = true;
        firePropertyChange(IWorkbenchPartConstants.PROP_DIRTY);
      }
    }
  }
```

Notice that `ContactEditor` extends `EditorPart` instead of `ViewPart`, just like `ContactListView` did. It also contains a `createPartControl` method that must be implemented. Our implementation creates SWT elements that compose a basic form for the contact's fields. It also fills these fields if the current editor input is not null, which will happen when we edit a contact instead of creating a new one. There's also an inline class named `ContactEditorModifyListener`, which, as the name implies, gets notified whenever a text element is modified. We use this class to change the `dirty` status to `true`. The `dirty` status means that the editor's input contains unsaved changes. This state is represented by an asterisk character (`*`) in front of the editor's name. This state will also allow us to enable or disable the **save** button accordingly. Notice how we change the `dirty` status in the `doSave` method.

Now, let's register the `ContactEditor` class in the `plugin.xml` file. Add a new extension for the `org.eclipse.ui.editors` extension point and add an editor element with the properties shown in the following screenshot:

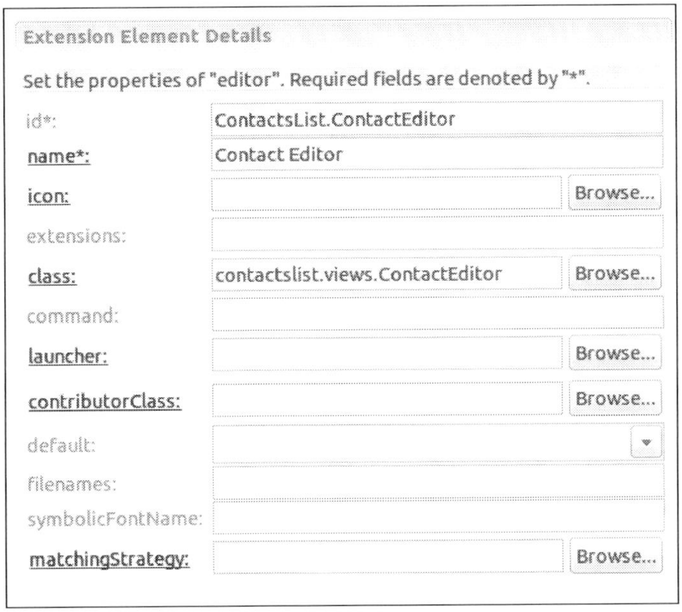

Editors are not added to perspectives as regular views are. Editors are opened programmatically by calling the `openEditor` method of the currently active `WorkbenchPage`, as we will see later in this chapter. They can be opened in any perspective, and they are placed in the editor space, which lies in the middle of the window.

The Save and New commands and menu entries

The next step is to implement commands, handlers, and menu entries that enable users to create new contacts and to save their work. Let's start with the handler classes. This is the code for the `save` command handler:

```
public class SaveContactHandler extends AbstractHandler {

  @Override
  public Object execute(ExecutionEvent event) throws
ExecutionException {
    PlatformUI.getWorkbench().getActiveWorkbenchWindow().
getActivePage().getActiveEditor().doSave(new NullProgressMonitor());

    return null;
  }
}
```

Note that there are two `AbstractHandler` classes, one in the `org.eclipse.core.commands` package and the other in `org.eclipse.ui.commands`. We'll be using the former because the latter is deprecated.

This is the new contact handler:

```
public class NewContactHandler extends AbstractHandler {

  @Override
  public Object execute(ExecutionEvent event) throws
ExecutionException {
    Contact contact = new Contact("New Contact", "", "");
    IViewPart view = PlatformUI.getWorkbench().
getActiveWorkbenchWindow().getActivePage().findView("ContactsList.
ContactsListView");
    ((ContactListView) view).addEntry(contact);
    return null;
  }
}
```

The new contact handler simply creates a new `Contact` object and calls the `addEntry` method in the `ContactsListView` view.

Let's proceed with the necessary modifications in the `plugin.xml` file, which is similar to what was demonstrated in the previous chapter. Add an extension to the `org.eclipse.ui.commands` extension point and add command elements for the `Save` and `New` commands. In the example, we use `contactsList.save` and `contactsList.new` as IDs and `Save` and `New` as names. Now add the `org.eclipse.ui.handlers` extension, and create two handler elements for the commands, linking them to the handler classes we just created.

The next step is to add the menu entries for these commands. Let's add a **File** entry in the main menu with the `New` and `Save` entries and a **Save** button in the toolbar. Add the elements shown in the following screenshot to create them:

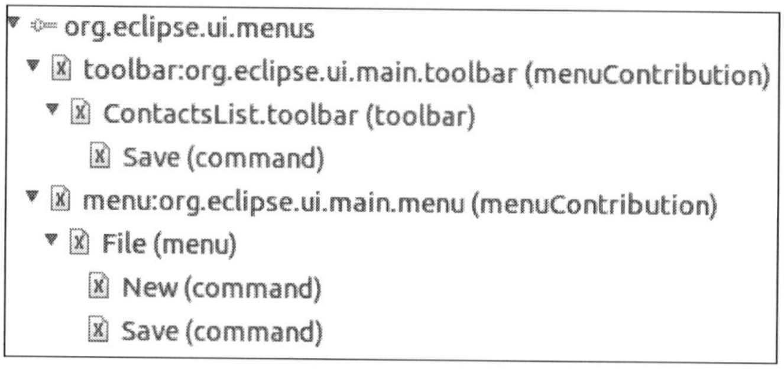

The `New` and `Save` commands point to the `contactsList.new` and `contactsList.save` command IDs respectively.

If you run the application at this point, you will realize that the **Save** button in the toolbar is not there. In fact, there's no toolbar in our application! That's because the implementation of the `ApplicationWorkbenchWindowAdvisor` extension provided by the template disables the toolbar. Open this class, and either comment out the `configurer.setShowCoolBar(false)` call inside the `preWindowOpen` method or change the parameter to `true` to display the toolbar.

Tying up the two views

Now that we have the two views and the menu entries, the next step is to connect the two views by opening an editor when the user double-clicks on an entry in the list. We will do that by adding a double-click listener to the `ContactList` view. Let's start by coding the listener. Create a class named `ContactListDoubleClickListener` with the following code:

```
public class ContactListDoubleClickListener implements
IDoubleClickListener {

  @Override
  public void doubleClick(DoubleClickEvent event) {
  TableViewer viewer = (TableViewer) event.getSource();
    StructuredSelection selection = (StructuredSelection) viewer.
getSelection();
    Contact contact = (Contact) selection.getFirstElement();
    try {
      PlatformUI.getWorkbench().getActiveWorkbenchWindow().
getActivePage().openEditor(new ContactEditorInput(contact),
"ContactsList.ContactEditor");
    } catch (PartInitException e) {
      e.printStackTrace();
    }
  }
}
```

When a `IDoubleClickListener` implementation is registered to a SWT widget, its `doubleClick` method is executed when the widget is double-clicked. Our `doubleClick` implementation gets the `TableViewer` widget and discovers which element was selected. This is another advantage of using content providers; they allow you to access the object to which the `TableViewer` entry refers easily.

Now that we have a double-click listener, add it to the contact's `TableViewer` widget by modifying the `createControl` method of `ContactListView` by adding the following line:

```
contactsViewer.addDoubleClickListener(new
ContactListDoubleClickListener());
```

To further integrate the contact's **view** window and the editor, we must forcibly refresh the contact's entries every time the editor is saved, in order to reflect a possible change in the contact's name. To do this, the `ContactsList` view will listen to a specific event that the editor will fire whenever it is saved. Let's start by adding a private `IViewPart` field named `contactListView` to the `ContactEditor` class that will contain the instance of the contacts view. Add the following line to the `createPartControl()` method of `ContactEditor` to get the instance:

```
contactListView = PlatformUI.getWorkbench().
getActiveWorkbenchWindow().getActivePage().findView("ContactsList.
ContactsListView");
```

Now add the following code to the end of the `doSave()` method to notify the contact list's instance that the editor was saved:

```
if(contactListView != null){
  if (contactListView instanceof IPropertyChangeListener){
    PropertyChangeEvent changeDirty = new PropertyChangeEvent(this,
"isDirty", true, false);
    ((IPropertyChangeListener)contactListView).
propertyChange(changeDirty);
  }
}
```

Now all we have to do is change the `ContactListView` class to implement the `IPropertyChangeListener` interface and add a `propertyChange()` method that acts accordingly when the editor is saved. Implement the `propertyChange` method with the following code:

```
@Override
public void propertyChange(PropertyChangeEvent event) {
  refreshEntries();
}
```

Running the application

We now have everything we proposed for this simple application. Let's see what it's looking like by running it. You should get a window like the following screenshot:

Navigate to **File** | **New** in the application's main menu to create a new contact. The editor will open. The following screenshot shows the editor:

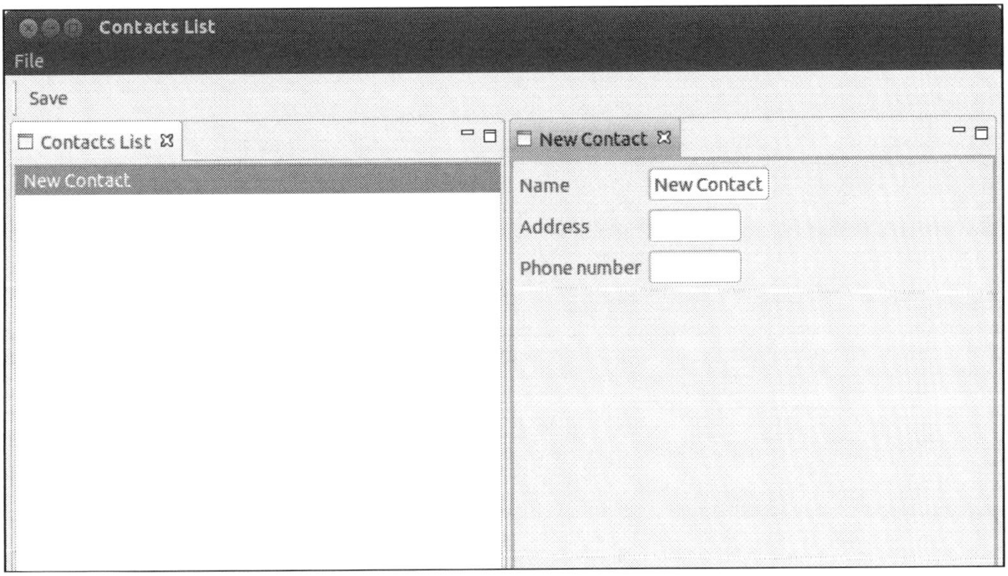

Packaging and branding an Eclipse RCP application

Our contact list application is up and running, but until now we can only run it from our development environment. In this section, we will see how we can transform our application into a self-contained product that can be distributed to users.

In order to achieve this, we need to create a product configuration file that describes how our product will be packaged and what it will contain. Fortunately, just like with the other plugin development configuration files, Eclipse has a wizard and a special editor for the product configuration file. To create the file using the wizard, right-click on your RCP application project in the `PackageExplorer` view, and navigate to **New | Product Configuration**. Make sure you are utilizing the **Plug-in Development** perspective in order to see these options. In the wizard window, type in `contactlist.product` as the product configuration filename. The wizard allows you to use an existing product or launch a configuration as the base configuration for the product. Because we have already launched the product inside our development environment, we already have a launch configuration that fits our basic needs. Select **Use a launch configuration** and select the `ContactsList` configuration. When you click on **Finish**, the product configuration editor will open, as shown in the following screenshot:

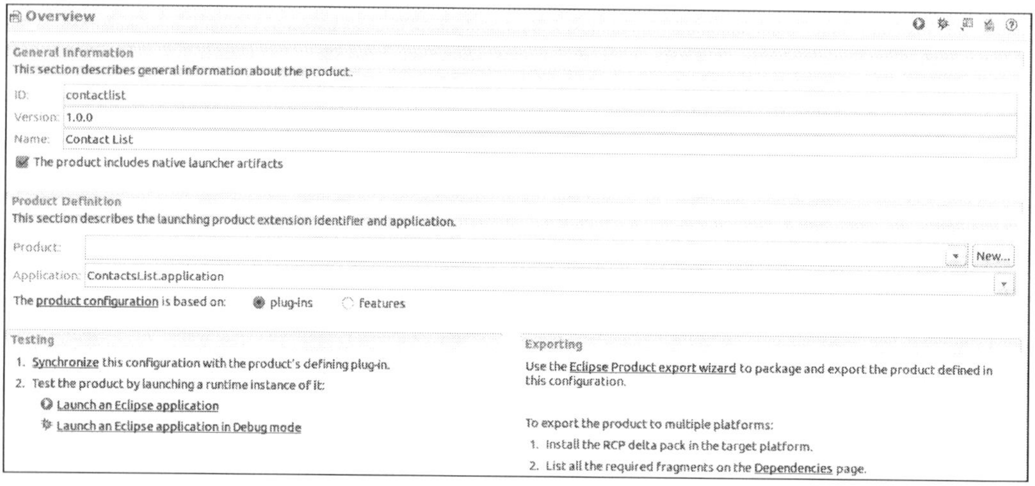

Fill the **General Information** section with the same values as the ones shown in the previous screenshot. In the **Product Definition** section, let's create a new product by clicking on the **New** button. Click on **Browse** to select `ContactList` as the defining plugin of the product. Leave **Product** as the product ID. In **Application**, select **ContactList.application**. Click on **finish**. We haven't used features in our product, so leave **plug-ins** selected as the product configuration.

Switch to the **Dependencies** tab of the product configuration editor. The list should already contain a number of plugins; `ContactsList` among them. This information came from the launch configuration we pointed to the wizard. You can check the list and add/remove plugins if necessary. Let's leave the ones that came from the launch configuration.

This should be enough for us to generate a bundle that contains our contact list application. Save the product configuration editor, right-click on the `ContactsList` project in the **Package Explorer** window, and select **Export**. Under the **Plug-in Development** category of the **Export** wizard, select **Eclipse product** and click on **Next**.

In the next page, point to the `contactlist.product` configuration file we just created in the **Configuration** field. The **Root directory** field determines the name of the folder that will contain our product. Because we want to generate a single file, select **Archive file** in the **Destination** field and point to a valid filename in your system. The `.zip` extension will be appended to the path you insert here. We don't want to include the source files in our product, so uncheck the **Export source** box. Your wizard page should look like the following screenshot:

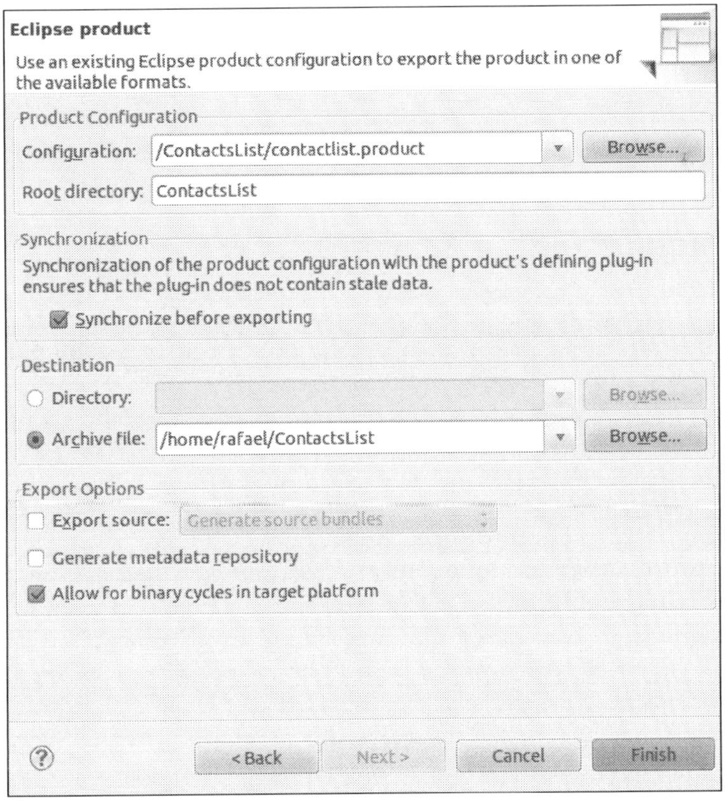

Click on **Finish** to start generating the product. When done, browse to the location you have pointed to in the wizard. You will find a `ContactsList.zip` file. Extract it and you will find an Eclipse-executable `eclipse.ini` configuration file and folders named `configuration` and `plugins`. As you can imagine, the `plugins` folder contains all the plugins added in the product configuration editor, including the `ContactsList` plugin. Double-click on the eclipse file and you will find our contact list application running outside of the development environment.

Branding an Eclipse RCP application

You have probably found it odd to have an executable named `eclipse` for an application called `ContactList`. The lack of icons and splash screens would also be a problem if this was an actual product. Let's fix these issues. Go back to the product configuration editor by double-clicking on the `contactlist.product` file. In the **Launching** tab, you will find a **Program Launcher** section. This is where you personalize your launcher. Select a proper name for our application. If you wish to add a launcher icon, add it to the `project` folder (the RCP application template generates an `Icons` folder; you can put the launcher icon there), and point to it in the **Icon** field. You can also specify launching arguments for both the application and the Java virtual machine in the **Launching** tab. Note that these configurations are specific to each platform (Linux, Mac OS X, Solaris, and Win32).

The **Splash** tab allows you to add a splash screen in your application. Splash screens are a useful feature for applications because they let users know that work is actually being done during the application's startup and it hasn't hung. To add a splash screen, one of your application's plugins must contain the image that will be used as the splash screen and as a bitmap file named `splash.bmp` in the plugin's root folder. Make sure the `splash.bmp` file is being included in the plugin's build by opening the **Plug-in Manifest Editor** window, switching to the **Build** tab, and ensuring the `splash.bmp` entry in the **Binary Build** list is checked. Once you have the splash screen file in place and properly included in the **Build** tab, just refer to the plugin in the **Location** section of the **Splash** tab. You can also include a progress bar or messages telling which part of the application is currently being loaded, similar to the ones in Eclipse.

The **Branding** tab allows selecting application window images. These images will be used in your operating system's toolbars and taskbars to represent your application. You can also customize the **About Dialog** message and image in this tab. To display the **About** window, you have to create an entry in the **applications** menu. because the **About** window still hasn't been ported to the command framework, we have to create the entry in the `ApplicationActionBarAdvisor` class. The following code shows an example of how the class can be implemented in order to create a **Help** menu with an **About Contact List** entry that displays the message and the image indicated in the `contactlist.product` file. The imports section is omitted.

```
public class ApplicationActionBarAdvisor extends ActionBarAdvisor {

    private IWorkbenchAction aboutAction;

  public ApplicationActionBarAdvisor(IActionBarConfigurer configurer)
{
        super(configurer);
    }

    protected void makeActions(IWorkbenchWindow window) {
      aboutAction = ActionFactory.ABOUT.create(window);
    }

    protected void fillMenuBar(IMenuManager menuBar) {
      MenuManager helpMenu = new MenuManager("&Help",
IWorkbenchActionConstants.M_HELP);
      menuBar.add(helpMenu);
      helpMenu.add(aboutAction);
    }
}
```

You can also add a welcome page to your application in the **Branding** tab, similar to the one that appears when you start Eclipse for the first time. By clicking on the **New** button in front of the **Intro ID** field, you will be presented with a welcome page wizard window. Select the current plugin as **Target**, select an ID for **Intro**, and click on **Finish**. This will create implementations for the `org.eclipse.ui.intro` and `org.eclipse.ui.intro.config` extension points in order to add the welcome page to your application. The welcome page's contents are described in an XML file placed in the plugin's root, named `introContent.xml`. Unfortunately, there's no specific editor for this file, so you will have to create and edit it manually.

> The `introContent.xml` format's contents are outside the scope of this book, but you can find plenty of information about it in the *Platform Plug-in Developer Guide* article by going to **User assistance support** | **Welcome** at `help.eclipse.org`.

Last but not least, the **Licensing** tab allows you to specify a license text and URL for your application.

Summary

In this chapter we have seen how to create a client application utilizing all the infrastructure Eclipse provides. By building a simple example application using the Eclipse RCP, we saw how easy it is to code and bundle a modular client application using SWT and JFace graphical elements.

Keyboard Shortcuts

Eclipse contains a huge number of keyboard shortcuts that, when mastered, can speed up your coding considerably. By using them, you avoid moving your hands from the keyboard to use the mouse. Using the mouse too frequently to perform actions can hurt your productivity.

Evidently, the idea is not to memorize the whole list in order to be super-productive. This list should be used as a reference for those moments when you realize that you have to perform a time-consuming task over and over, and you ask yourself, "Is there a faster way of doing this?"

This is not an extensive list. We have selected some of the most useful shortcuts and organized them in topics to make it easier to browse.

File editor shortcuts

In this section, we'll present some shortcuts that can be useful when using file editors. Mastering them can ease tasks such as code editing, navigation, and code generation.

Code edition shortcuts

The following table contains some of the most used shortcuts related to code editing:

Shortcut	**Purpose**
Ctrl + C	Copy
Ctrl + X	Cut
Ctrl + V	Paste
Ctrl + Z	Undo
Ctrl + Y	Redo

Shortcut	Purpose
Ctrl + *F*	Find/replace
Ctrl + *D*	Deletes the current line
Alt + up arrow	Moves the current line up or down
Alt + down arrow	
Ctrl + *Shift* + *Delete*	Deletes the current line's content after the caret
Ctrl + *Shift* + *C*	Comments/uncomments the currently selected lines by adding "//"
Ctrl + *Shift* + /	Comments the current selection by adding /* ... */
Alt + *Shift* + *J*	Generates element comment to the currently selected element
Ctrl + *Shift* + *B*	Adds/removes breakpoint in the current line
Tab	Increases/decreases the selected lines' indentation
Shift + *Tab*	
Ctrl + *I*	Fixes the selected lines' indentation
Ctrl + *Shift* + up arrow	Expands the current selection to enclosing element (selects whole method if a portion of the method's code is selected, for example)
Ctrl + *Shift* + left arrow	Expands the current selection to next/previous element
Ctrl + *Shift* + right arrow	

Code generation and code refactoring shortcuts

As we have seen in *Chapter 2, Java Development*, Eclipse can generate and refactor code for you. It can perform tasks such as creating method snippets, organizing a class' imports, among others. The following table shows some shortcuts for these tasks:

Shortcut	Purpose
Alt + *Shift* + *R*	Renames the currently selected method
Alt + *Shift* + *M*	Extracts the current selection to a method
Alt + *Shift* + *O*	Organizes imports
Alt + *Shift* + *F*	Formats the selection
Alt + *Shift* + *S*	Opens the Source dialog (allows generating getters/setters, constructors, to String methods, and so on)
Ctrl + 1	Opens the Quick Fix dialog

Code navigation shortcuts

In big projects, navigating through your source code can be a painful task. Big and numerous source files make it hard to find what you want. Having the following shortcuts under your sleeve might be handy:

Shortcut	Purpose
Ctrl + *Shift* + up arrow *Ctrl* + *Shift* + down arrow	Moves to next/previous element of the source code (parameter, method, and so on)
Ctrl + *Shift* + *P*	When a bracket is selected, it goes to matching bracket
Ctrl + *L*	Goes to a line
Ctrl + *Q*	Goes back to last edit location
Ctrl + . *Ctrl* + ,	Goes to next/previous annotation (warnings, errors, and so on)
Ctrl + *Page Up* *Ctrl* + *Page Down*	Goes to next/previous editor window
Ctrl + left-click	When a class name is selected, opens its source code (if available in workspace)
Ctrl + *Shift* + *T*	Open Type - Allows you to search workspace's classes by name
Ctrl + *Shift* + *R*	Open the Resource—Allows you to search workspace's resources by name

Java shortcuts

The following Java-specific shortcuts are also very useful:

Shortcut	Purpose
F2	Opens the currently selected class' JavaDoc in a tooltip box
F3	Goes to the currently selected element definition (class definition if class is selected, variable declaration if variable is selected)
F4	Opens the Type Hierarchy view for the currently selected class
Ctrl + O	Opens the outline tooltip box for the class currently being edited
Ctrl + Alt + H	Opens the Call Hierarchy view for the method currently being edited

File shortcuts

In the following table, you will find some shortcuts for tasks regarding files (saving, opening, and so on):

Shortcut	Purpose
Ctrl + S	Saves the current file
Ctrl + Shift + S	Saves all the open files
Ctrl + W	Closes the current editor window
Ctrl + Shift + W	Closes all the editor windows
Ctrl + N	Opens the New File Wizard

Run and debug shortcuts

The following shortcuts can make running and debugging projects faster and easier:

Shortcut	Purpose
Ctrl + F11	Runs the current project
F11	Debugs the current project
F5	Step into (in debug mode)
F6	Step over (in debug mode)
F7	Step return (in debug mode)
F8	Resume (in debug mode)

Views shortcuts

Some views can be opened or displayed using shortcuts. Just use the *Alt* + *Shift* + *Q* shortcut, then press the letter corresponding to the view you want to open. If you don't press any keys after the *Alt* + *Shift* + *Q* shortcut, a list will appear with the view's letters. The following table contains the letters for the most used views:

View	**Letter**
Breakpoints	*B*
Console	*C*
Outline	*O*
Package Explorer	*P*
Problems	*X*
Type Hierarchy	*T*

Index

Symbols

7-Zip
  about 11
  URL, for downloading 11
@Before annotation 61
@BeforeClass annotation 61
@Ignore annotation
  creating 58
@Test annotation 55

A

Abstract Windowing Toolkit. *See* **AWT**
actions
  versus commands 181
Activator.java class 170
Apache Hadoop repository 78
Apache Software Foundation 130
API (Application Programming Interface) 199
ArmEvent 110
assertEquals() method 59
AWT 87

B

BasicCalculator class 51
benefits, Java web application technologies 128
build.properties file 169
button widgets 96, 97

C

calculator example
  about 152, 153
  deploying 154-159
Call Hierarchy view 32
category 190
client application
  developing, Eclipse RCP used 201
client application project
  creating, template used 202
code
  editing 37
  generating 34
code navigation, Java Editor view 26, 27
code refactoring
  about 38
  signature, modifying for method 39
  variables, renaming 38
ColorDialog 124
combo widgets 99, 100
commands
  versus actions 181
comment templates
  editing 37, 38
commit option 84
Common Gateway Interface (CGI) 128
Concurrent Versioning System (CVS) 73
config.ini file 172
constructors
  generating 35
contact editor 213-216

contact list application
contact editor 213-216
contact list view 210-212
creating 208, 209
menu entries 217, 219
New command 217, 219
running 221
Save command 217, 219
views, tying up 219
contact list view 210-212
content assist, Java Editor view 25, 26
context help
adding, to view 191
contributions
disabling 184-187
restricting 184-187
ControlEvent 110
controls
about 92
button widgets 96, 97
combo widgets 99, 100
label widgets 92, 93
list widgets 97, 98, 99
text widgets 94, 95
CreateWorkbenchWindowAdvisor method 204
CVS
about 23, 74
Branch option 76
project, checking out 76
synchronization feature 76
CVS repositories 74, 75

D

debugger
using 67-71
debugging 43
destroy() method 140
dialogs
about 121
ColorDialog 124
FileDialog 123
FontDialog 125
MessageBox 121, 122
PrintDialog 126
Display class 92, 109
DisposeEvent 110
doGet method 153
drop down menus 115

E

Eclipse
about 10, 43, 197
downloading 10
installing 11
URL, for downloading 10
Eclipse Juno
overview 11-18
Eclipse RCP
about 197, 198
Eclipse workbench 201
JFace 200
OSGi framework implementation 199
SWT 200
used, for developing client application 201
Eclipse RCP application
branding 224, 225
packaging 222, 223
Eclipse Web Tools Platform (WTP)
about 134, 136
server, configuring 137-140
URL 134
URL, for downloading 134
used, for implementing Java servlet 141-151
used, for implementing JSP 160-163
Eclipse workbench 201
EGit 80, 82
equals() method
generating 36
Equinox 199
errors, Java Editor view
compiling 25
events 109
eXtensible Markup Language. *See* **XML**
extension points
about 173
declaring 173-175
implementation, using in code 177, 178
implementing 179, 180
interfaces, providing 176, 177

F

fail() method 55
FastCGI 128
FileDialog 123
file editor shortcuts
about 227
code edition shortcuts 227, 228
code generation shortcuts 228
code navigation shortcuts 229
code refactoring shortcuts 228
Java-specific shortcuts 229
file shortcuts 230
FillLayout class
about 101
fields 101
FIRST principles, unit testing 44
FocusEvent 111
FontDialog 125
Formatter view 34
full-blown debugger 67

G

generated source code 203-206
getData method 91
getInitialWindowPerspectiveId method 204
getSelection method 97
getters
about 34
generating 34, 35
Git
about 80
versus Subversion (SVN) 80
Git repository
creating 82
using 83
Graphical User Interface (GUI) 87, 109
GridLayout class
about 104
fields 104-106

H

hashCode() method
generating 36
HelpEvent 111
HTML 128, 129
HTTP 128
HttpServlet class 129
HyperText Markup Language. *See* **HTML**
HyperText Transfer Protocol. *See* **HTTP**

I

init() method 140
installation, Eclipse 11
installation, Tomcat 130-133
Integrated Development Environment (IDE) 9, 19, 197

J

Java 128
Java class
creating 21
Java Development Kit (JDK) 10
Java Development Tools (JDT) 45, 206
Java Editor view
about 24
code navigation 26, 27
content assist 25, 26
customizing 29, 30
errors, compiling 25
Quick fix 27-29
warnings annotations, compiling 25
Java project
creating 20
importing 22, 23
importing, from Version Control Servers 23
Java class, creating 21
working sets, creating 21, 22
Java Runtime Environment (JRE) 10
JavaServer Pages (JSP)
about 127, 129
implementing, Eclipse WTP used 160-163
Java servlet
about 127
implementing, Eclipse WTP used 141-151
Java views
about 23
Call Hierarchy view 32
Formatter view 34
Java Editor view 24
Organize Imports view 33
Outline view 31

Package Explorer view 23
Problems view 30
Save Actions view 33
Type Hierarchy view 32
Java Virtual Machine (JVM) 10, 87
Java web application technologies
benefits 128
javax.servlet 129
javax.servlet.http package 129
JFace 120, 200
JUnit
about 44, 45
setting up 46-48
URL 45
used, for testing 49-52
JUnit 3
versus JUnit 4 47
JUnit 4
versus JUnit 3 47
junit.jar file 46
JUnit Runner view
overview 57, 58
Juno version 77
JVM 130

K

keyboard shortcuts
file editor shortcuts 227
file shortcuts 230
run and debug shortcuts 230
views shortcuts 231
KeyEvent 111, 112

L

label widgets 92, 93
launch configuration
creating 40, 41
layout classes
about 100
FillLayout 101, 102
GridLayout 104-106
RowLayout 102, 104
layouts 100
life cycle layer 199
Listener 110
list widgets 97-99

M

main menus 115
MANIFEST.MF file 169
menu entries 217, 219
MenuEvent 111
Menu events 115
MenuItem events 116
menus
about 115
drop down 115
main menu 115
pop up 115
MessageBox dialog 121, 122
method comments
generating 36
ModifyEvent 111
mod_perl 128
module layer 199
MouseButtonPressed event 114
MouseButtonReleased event 114
MouseDoubleClick event 114
MouseEvent 111, 114, 115
MouseMoveEvent 111
MouseTrackEvent 111

N

New command 217, 219
NSAPI 128

O

Organize Imports view 33
org.eclipse.swt.events package 110
org.eclipse.ui.commands 182
org.eclipse.ui.handlers 182, 183
org.eclipse.ui.menus 181
OSGi framework 199
OSGi framework implementation, Eclipse RCP
life cycle layer 199
module layer 199
security layer 200
service layer 200
Outline view 31

P

Package Explorer view 23
PaintEvent 111
plugin
creating 168-170
exporting 194
Plugin Manifest Editor 170
plugin project
configurations, debugging 171, 172
configurations, running 171, 172
debugging 171
running 171
Plugin Spy feature 188, 189
plugin.xml file 169, 173
pop up menus 115
PrintDialog 126
Problems view
about 30
customizing 30, 31
project
building 39
launch configuration, creating 40, 41
running 39
project build path
managing 41, 42

Q

Quick fix, Java Editor view 27-29

R

RCP 197, 198
RCP applications
debugging 207
running 207
readAndDispatch method 92, 109
Rich Client Platform. *See* **RCP**
RowLayout class
about 102
fields 102
run and debug shortcuts 230

S

Save Actions view 33
Save command 217, 219
schema editor
example 174
security layer 200
SelectionEvent 111
service layer 200
service() method 140
servlet lifecycle 140
servlets 129
setAppName method 92
setData method 91
setLocation method 100
setSelection method 97
setSize method 100
setters
about 34
generating 34, 35
shell 91
ShellEvent 111
source code management 73
Standard Widget Toolkit. *See* **SWT**
StickyView 190
StyledText widget 107
Subclipse project 77
Subversion (SVN)
about 73-80
versus Git 80
Subversive project 77
SWT
about 87, 88, 200
setting up 88-90

T

table 118, 120
teardown() method 61
template
used, for creating client application project 202, 203
testAdd() method 57
test cases
creating 58-66

Test Driven Development (TDD) 44
testing
with JUnit 49-52
test methods
implementing 53-57
testSubtract() method 60
text widgets 94, 95
Tomcat
about 127, 130
installing 130-133
toolbar 117, 118
ToolBar styles 117
ToolItem events 118
ToolItem styles 117
toString() method
generating 36
trace code 67
TraverseEvent 111
TreeCursor class 107
TreeEvent 111
typed events 110
Type Hierarchy view 32

U

unit testing
about 44
FIRST principles 44
untyped events 110

V

VerifyEvent 111
version control 73
Version Control Servers
Java project, importing from 23
version control systems (VCS)
about 73
CVS 74
Git 80-85
SVN 77-80
viewers 120
views
context help, adding 191
creating 190
current state, saving 191
example 192-194
views shortcuts 231

W

WAR files
generating 164
warnings annotations, Java Editor view
compiling 25
Web application Archive. *See* **WAR files**
widgets 90, 91
working sets
creating 21, 22

X

XML 129

[PACKT] open source
PUBLISHING
community experience distilled

Thank you for buying

Getting Started with Eclipse Juno

About Packt Publishing

Packt, pronounced 'packed', published its first book "*Mastering phpMyAdmin for Effective MySQL Management*" in April 2004 and subsequently continued to specialize in publishing highly focused books on specific technologies and solutions.

Our books and publications share the experiences of your fellow IT professionals in adapting and customizing today's systems, applications, and frameworks. Our solution based books give you the knowledge and power to customize the software and technologies you're using to get the job done. Packt books are more specific and less general than the IT books you have seen in the past. Our unique business model allows us to bring you more focused information, giving you more of what you need to know, and less of what you don't.

Packt is a modern, yet unique publishing company, which focuses on producing quality, cutting-edge books for communities of developers, administrators, and newbies alike. For more information, please visit our website: `www.packtpub.com`.

About Packt Open Source

In 2010, Packt launched two new brands, Packt Open Source and Packt Enterprise, in order to continue its focus on specialization. This book is part of the Packt Open Source brand, home to books published on software built around Open Source licences, and offering information to anybody from advanced developers to budding web designers. The Open Source brand also runs Packt's Open Source Royalty Scheme, by which Packt gives a royalty to each Open Source project about whose software a book is sold.

Writing for Packt

We welcome all inquiries from people who are interested in authoring. Book proposals should be sent to author@packtpub.com. If your book idea is still at an early stage and you would like to discuss it first before writing a formal book proposal, contact us; one of our commissioning editors will get in touch with you.

We're not just looking for published authors; if you have strong technical skills but no writing experience, our experienced editors can help you develop a writing career, or simply get some additional reward for your expertise.

[PACKT] open source
community experience distilled
PUBLISHING

Eclipse 4 Plug-in Development by Example Beginner's Guide

ISBN: 978-1-782160-32-8 Paperback: 348 pages

How to develop, build, test, package, and release Eclipse plug-ins with features for Eclipse 3.x and Eclipse 4.x

1. Create plug-ins to extend the Eclipse runtime covering Eclipse 3.x and the changes required for Eclipse 4.x
2. Plug-ins from design to distribution – wide coverage of the entire process
3. No prior OSGi or Eclipse plug-in development experience necessary

---

Java EE Development with Eclipse

ISBN: 978-1-782160-96-0 Paperback: 426 pages

Develop Java EE applications with Eclipse and commonly used technologies and frameworks

1. Each chapter includes an end-to-end sample application
2. Develop applications with some of the commonly used technologies using the project facets in Eclipse 3.7.
3. Clear explanations enriched with the necessary screenshots

Please check **www.PacktPub.com** for information on our titles

26481971R00138

Made in the USA
Lexington, KY
08 October 2013